Atlantis Briefs in Artificial Intelligence

Volume 1

More information about this series at htttp://www.atlantis-press.com

Nan Cao · Weiwei Cui

Introduction to Text Visualization

ATLANTIS
PRESS

Nan Cao
IBM T. J. Watson Research Center
Yorktown Heights, NY
USA

Weiwei Cui
Microsoft Research Asia
Beijing
China

Atlantis Briefs in Artificial Intelligence
ISBN 978-94-6239-185-7   ISBN 978-94-6239-186-4   (eBook)
DOI 10.2991/978-94-6239-186-4

Library of Congress Control Number: 2016950403

Printed on acid-free paper

Acknowledgements

We would like to thank Prof. Yu-Ru Lin from University of Pittsburgh for her initial efforts on discussing the outline and the content of this book. We also would like to thank Prof. Qiang Yang from the Hong Kong University of Science and Technology who invited us to write the book.

Contents

Chapter 1
Introduction

Abstract Text is one of the greatest inventions in our history and is a major approach to recording information and knowledge, enabling easy information sharing across both space and time. For example, the study of ancient documents and books are still a main approach for us to studying the history and gaining knowledge from our predecessors. The invention of the Internet at the end of the last century significantly speed up the production of the text data. Currently, millions of websites are generating extraordinary amount of online text data everyday. For example, Facebook, the world's largest social media platform, with the help of over 1 billion monthly active users, is producing billions of posting messages everyday. The explosion of the data makes seeking information and understanding it difficult. Text visualization techniques can be helpful for addressing these problems. In particular, various visualizations have been designed for showing the similarity of text documents, revealing and summarizing text content, showing sentiments and emotions derived from the text data, and helping with big text data exploration. This book provides a systematical review of existing text visualization techniques developed for these purposes. Before getting into the review details, in this chapter we introduce the background of information visualization and text visualization.

1.1 Information Visualization

In 1755, the French philosopher Denis Diderot made the following prophecy:

> As long as the centuries continue to unfold, the number of books will grow continually, and one can predict that a time will come when it will be almost as difficult to learn anything from books as from the direct study of the whole universe. It will be almost as convenient to search for some bit of truth concealed in nature as it will be to find it hidden away in an immense multitude of bound volumes. – Denis Diderot

About two and a half centuries later, this prophecy has come true. We are facing a situation of *Information Overload*, which refers to the difficulty a person may have in understanding an issue and making decisions because of the presence of too much information. However, *Information Overload* is not mainly caused by the growth of books but mainly by the advent of the Internet.

© Atlantis Press and the author(s) 2016
C. Nan and W. Cui, *Introduction to Text Visualization*, Atlantis Briefs
in Artificial Intelligence 1, DOI 10.2991/978-94-6239-186-4_1

Several reasons could be cited for the Internet accelerating the process of information overload process. First, with the Internet, the generation, duplication, and transmission of information has never been easier. Blogging, Twitter, and Facebook provide ordinary people the ability to efficiently produce information, which could be instantaneously accessed by the whole world. More and more people are considered active writers and viewers because of their participation. With the contribution of users, the volume of Internet data has become enormous. For example, 161 exabytes of information were created or replicated in the Internet in 2006, which were already more than that the generated information in the past 5000 years [6]. In addition, the information on the Internet is constantly updated. For example, news websites publish new articles even every few minute; Twitter users post millions of tweets every day, and old information hardly leaves the Internet. For this kind of huge amount of information, analysis requires digging through historical data, which clearly complicates understanding and decision making. Furthermore, information on the Internet is usually uncontrolled, which likely causes high noise ratio, contradictions, and inaccuracies in available information on the Internet. Bad information quality will also disorientate people, thereby causing the information overload.

Understanding patterns in a large amount of data is a difficult task. Sophisticated technologies have been explored to address such an issue. The entire research field of data mining and knowledge discovery are dedicated to extracting useful information from large datasets or databases [5], for which data analysis tasks are usually performed entirely by computers. The end users, on the other hand, are normally not involved in the analysis process and passively accept the results provided by computers.

These issues could be addressed via *information visualization* techniques whose primary goal is to assist users *see* information, *explore* data, *understand* insightful data patterns, and finally supervise the analysis procedure. Research in this filed are motivated by the study of perceptions in psychology. Scientists have shown that our brains are capable of effectively processing huge amounts of information and signals in a parallel way when they are properly visually represented. By turning huge and abstract data, such as demographic data, social networks, and document corpora, into visual representations, information visualization techniques help users discover patterns buried inside the data or verify the analysis results.

Various definitions of *information visualization* exist [1, 3, 7] in current literature. One of the most commonly adopted definitions is that of Card et al. [2]: "the use of computer-supported, interactive visual representations of abstract to amplify cognition". This definition highlights how visualization techniques help with data analysis, i.e., the computer roughly processes the data and displays one or some visual representations; we, the end users, perform the actual data analysis by interacting with the representations.

A good visualization design is able to convey a large amount of information with minimal cognitive effort. Considered as a major advantage of visualization techniques, this feature is informally described by the old saying "A picture is worth a thousand words".[1] Communicating visually is more effective than using text based on the following reasons:

- The human brain can process several visual features, such as curvature and color, much faster than symbolic information [12]. For example, we can easily identify a blue dot in a large number of red dots, even before consciously noticing it.
- Information visualization takes advantage of the high bandwidth of the human perceptual system in rapidly processing huge amounts of data in a parallel way [13].
- Information visualization could change the nature of a task by providing external memory aids, and by providing information that could be directly perceived and used without being interpreted and formulated explicitly [14]. These external aids are described by Gestalt laws [8], which summarize how humans perceive visual patterns.

Another major advantage of visualization is "discovering the unexpected" [11, 13]. Normal data mining or knowledge discovery methods require a priory question or hypothesis before starting the analysis. Without any priory knowledge to a data, we will have to enumerate all possibilities, which is time-consuming and insecure. Meanwhile, information visualization is an ideal solution for starting the analysis without assumptions, and can facilitates the formation of new hypotheses [13].

A popular example for "discovering the unexpected" is Anscombe's quartet (see Fig. 1.1a), which consists of four sets of data. If we only look at the statistics that describe each of them (see Table 1.1), we may easily jump to the conclusion that these four datasets are very similar. However, this idea has been disproven. Figure 1.1b shows the scatter plots for each data set. Everyone can tell that these four datasets are not alike at all. Once people see the differences, they then may have a better idea on choosing the correct statistical metrics. In addition, information visualization requires less intense data analysis compared with data mining techniques. Therefore, by faithfully visualizing information in the way, the data are collected, instead of overcooking it, and data errors or artifacts may have higher chances to reveal themselves [13].

The development of computer hardware, particularly the human computer interface equipment and advanced graphics cards in the mid-1990s, has further facilitated people's exploration of global and local data features [13]. In 1996, Ben Shneiderman systematically presented the visual information seeking mantra [10] to design information visualization systems: "overview first, zoom and filter, then details-on-demand". This mantra categorizes a good visualization process into three steps, which allow users to explore datasets by using various level-of-detail visual representations. At the first step, visualization should provide the general information on entire datasets. Without holding assumptions, we could have a less-biased perspective of the entire data, understand its overall structure, and finally identify interesting

---

[1]This phrase "A picture is worth a thousand words" was first used by Fred R. Barnard in an advertisement entitled "One Look is Worth A Thousand Words." in 1921. In another advertisement designed by Barnard in 1927, he attached a Chinese proverb to make the ad more serious. Sadly, there is no evidence shows that the attached Chinese proverb really exists in Chinese literature.

(a)   **Anscombes quarte**

I		II		III		IV	
x	*y*	*x*	*y*	*x*	*y*	*x*	*y*
10.0	8.04	10.0	9.14	10.0	7.46	8.0	6.58
8.0	6.95	8.0	8.14	8.0	6.77	8.0	5.76
13.0	7.58	13.0	8.74	13.0	12.74	8.0	7.71
9.0	8.81	9.0	8.77	9.0	7.11	8.0	8.84
11.0	8.33	11.0	9.26	11.0	7.81	8.0	8.47
14.0	9.96	14.0	8.10	14.0	8.84	8.0	7.04
6.0	7.24	6.0	6.13	6.0	6.08	8.0	5.25
4.0	4.26	4.0	3.10	4.0	5.39	19.0	12.50
12.0	10.84	12.0	9.13	12.0	8.15	8.0	5.56
7.0	4.82	7.0	7.26	7.0	6.42	8.0	7.91
5.0	5.68	5.0	4.74	5.0	5.73	8.0	6.89

(b)

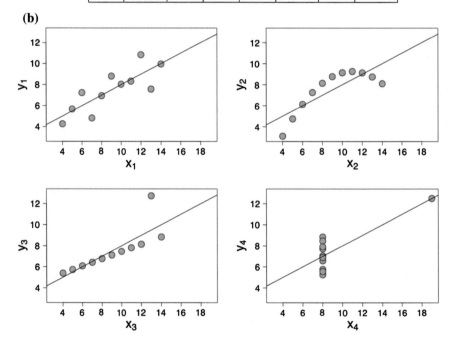

Fig. 1.1 Anscombe's quartet: **a** four different datasets; **b** scatter plots corresponding to each data set in (**a**)

areas to explore. At the second step of "zoom and filter", we isolate the areas of interest and generate reasonable hypotheses regarding patterns that draw our attentions. Finally, the details of the data should be displayed for further analysis, and eventually confirming or disprove our hypotheses. Given that details are extracted from an isolated area, we could avoid being overwhelmed by the large quantity of information

Table 1.1 Same statistics in Anscombe's quartet

Property (in each set)	Value
Mean of x	9.0
Variance of x	10.0
Mean of y	7.50
Variance of y	3.75
Correlation between x and y	0.898
Linear regression line	$y = 0.5x + 3.0$

and analyze the details more efficiently. Many visual analysis approaches follow the same mantra, offering general users an interactive and intuitive way to explore their data.

In addition, real world data usually have multiple data dimensions that show information from different aspects. For example, user profiles in a social network are usually multidimensional in describing the age, sex, and location of users. The description of product features is multidimensional; even an unstructured text corpora is multidimensional because of it consists of multiple topics. Increasing the data variables and dimensions leads to a more precise description of information. However, this feature significantly increases the difficulties of data analysis. Understanding a four dimensional dataset already exceeds the human capability of perception, let alone detects the multidimensional data patterns by analysis. This challenge has attracted great attentions in the past decades. Efforts have been made to reveal relational patterns such as correlations, co-occurrences, similarities, dissimilarities, and various sematic relations such as friendships in a social network and causalities of news events.

Many sophisticated data mining techniques, such as correlation analysis, clustering, classification, and association rule detection have been developed to detect relational patterns among data entities based on their attributes. For example, data entities within the same cluster or class show similarity of certain attributes. Although these techniques have been successfully applied to analyze large multidimensional datasets, such as documents and databases, they have critical limitations.

First, data mining techniques are designed to solve or verify predefined tasks or assumptions of datasets. These techniques always lead to expected analysis results. For example, the correlation of two data variables A and B could be tested by correlation analysis. A cluster analysis divides data entities into a predefined number of groups. These techniques are highly driven by user experiences in data analysis and their priori domain knowledge about the datasets. Therefore, a challenging problem that requires new techniques is the detection of unexpected data patterns of common users, who have minimal data analysis experiences, using limited prior knowledge.

Second, training sets or ground truths of a dataset are unavailable, which makes machine learning and evaluation difficult. Lacking training sets disables supervised learning, such as classification. Moreover, lacking of ground truth causes difficulty in evaluating the analysis results of unsupervised learning, such as cluster analysis. To explain this point effectively, we examine the detailed process of typical cluster analysis. Cluster analysis is a widely used analytical method to group data entities into subsets called clusters, such that the entities in each cluster are similar in a certain way.

In this process, users are only required to choose a distance function (e.g., Euclidean distance) that measures similarity of two data items in a feature space, as well as other parameters, such as the number of clusters or a maximum cluster diameter. These predefined parameters are critical in the analysis and challenging to decide. For example, users must provide the number of clusters (i.e., k) for the well-known K-means algorithm. However, selecting a proper k value is difficult when the underlying given data are unknown ground truth. Therefore, algorithms such as K-means might group together entities that are semantically different (when k is smaller than the real number of clusters) or separate entities that are semantically similar (when k is larger than the real number of clusters). Therefore, without a ground truth, correct evaluation of analysis result is difficult.

Third, the analysis results of multidimensional dataset could be difficult for users to interpret because of several reasons. First, datasets may contain heterogeneous dimensions with different types. The inconsistency of the dimensions makes the analysis results difficult to understand. For example, a challenging task is to understand the changing of topic trends in a text flow given that the process requires extracting textual and temporal information from the data. Second, interpretation is still difficult even if all dimensions are of the same type. For example, cluster analysis of multivariate datasets may generate results that are difficult to understand. Specifically, in multivariate clusters, data items are grouped together if they are close in the multidimensional feature space. However, their similarities may be mainly due to their closeness in a subset of dimensions instead of all dimensions. Understanding these abstract relationships is challenging. Moreover, a cluster may contain several different sub-clusters that could have different meanings for users. This sub-cluster structure is usually difficult to detect. Third, complete datasets are heterogeneous when collected from multiple data sources. For example, to analyze the process of information diffusion in microblogs, we need to collect social connections between users, critical events as triggers of diffusion, and places where information is spread from data sources, such as Twitter, online news, and Google maps. Understanding such dataset could be difficult given that all pieces of data need to be seamlessly assembled to provide unique and meaningful information.

Information visualization could be of great value when addressing these problems. From the very beginning, information visualization has been one of the most important techniques to facilitate interpretation, comparison, and inspection of analytical results as well as the underlying raw data. In the 1990s, information visualization techniques started to be used for analyzing multidimensional datasets. Techniques, such as the pixel-oriented database visualization designed by Keim and Kriegel [9, 10] and the parallel coordinates for data correlation analysis designed by Inselberg [7] were introduced to review simple relational patterns. These early visualizations have been extensively studied over the past years, and various novel visualizations with rich interactions are designed to represent the multidimensional datasets from different perspectives. Several visualizations focus on visualizing the relationships of individual items and their attributes while others focus on representing the overview of data distributions. Based on these interactive visualizations, explorative visual analysis becomes a popular approach for detecting relational data patterns. Visual analysis collects user feedback through interactions to refine the underlying analysis models. This step allows for highly correct results and precise analysis controls that could be achieved. Particularly, the motivation behind interactive multidimensional visualizations is to help with automatic data analysis for testing assumptions, selecting analysis factors, models, and estimators, as well as revealing data patterns and detecting outliers. To visualize intuitively the relationships inside a multidimensional dataset, three primary challenges have emerged.

The first challenge is the process of intuitively encoding data that contain multiple dimensions to reveal their innate relations. High dimensions add difficulty when visual primitives, such as nodes, and lines with visual features, such as colors, shapes, textures and positions, are limited. The limitation of such data makes encoding difficult. In addition, several visual features conflict with each other and could not be used simultaneously (e.g., color and texture). In addition, several features are unsuitable for precise representation of certain types of data (e.g., color and area could not be used to represent numerical values precisely). Thus, intuitive encoding is required in determining a set of visual primitives with proper visual attributes as well as encoding data dimensions and their relations while minimizing visual clutter. In certain situations, data are split by dimensions and visualized in different views, such as PaperLens [9] because of the difficulty involved in encoding multiple dimensions within single visualization. In this case, the challenge is the seamless connection of these views to reveal the different aspects of data while preserving user mental maps and reducing the training process.

The second challenge is detecting relational patterns in the multidimensional datasets based on their visual representations. These patterns could be complicated given their inclusion of substructures and correlations over various dimensions, clusters over data entities, similarities of clusters, and various semantic relationships across heterogeneous data pieces. Identifying various relationships could be difficult because raw datasets or analysis results are complex and difficult to represent. Moreover, automatic analysis may generate misleading results. Thus, visualizations should provide additional aid in representing the results and in error detection and evaluation.

Finally, the third challenge is the way interactions could refine analysis results and the underlying analysis models. For example, a critical problem occurs when the user states his/her constraints on visualization. Limited research has been conducted in this area. This challenge depends on new interaction techniques, and more importantly on several hybrid techniques, such as active learning and incremental mining.

The above challenges could be trickier when data are large and dimensions are high. In these situations, visual clutter, such as line crossings and node overlaps, are usually unavoidable. The computation performance could be another problem given that most visualization techniques require efficient computation to provide online layout and analysis for interaction to be conducted. The final and trickiest problem is the limitation of the cognitive capability of human beings. Therefore, users could understand how the visual representation of huge and high dimensional data becomes a problem. Therefore, many techniques have been designed to tackle these problems. For example, various visual clutter reduction methods [4] are introduced for different data types and visualizations. Statistical embedding techniques are designed to represent huge datasets based on statistical aggregation of individual data items. This design allows data overview to be clearly represented in chat applications. Dimension reduction methods, such as projection, are introduced to map high-dimensional information spaces into low-dimension plains. Several visualizations are carefully designed based on the aforementioned techniques to provide intuitiveness of huge and high-dimensional datasets, which are surveyed in the following section.

1.2   Text Visualization

Large collections of text documents have become ubiquitous in the digital age. In areas ranging from scholarly reviews of digital libraries to legal analysis of large email databases and online social media and news data, people are increasingly faced with the daunting task of needing to understand the content of unfamiliar documents. However, this task is made challenging because of the extremely large text corpus and dynamical change in data over time, as well as information coming from multiple information facets. For example, online publication records contain information on authors, references, publication dates, journals, and topics. The varied data cause challenges in understanding how documents relate with one another within or across different information facets. To address this issue, interactive visual analysis techniques that help visualize such content-wised information in an intuitive manner have been designed and developed. These techniques enable the discovery of actionable insights.

Existing text visualization techniques were largely designed to deal with the following three major forms of text data:

- **Documents**. A text document refers to the data, such as a paper, a news article, or an online webpage. Visualizations designed to represent documents usually focus on illustrating how various documents are related or on summarizing the content or linguistic features of a document to facilitate an effective understanding or comparison of various documents.
- **Corpus**. A corpus indicates a collection of documents. Visualizations designed to represent the corpus usually focus on revealing statistics, such as topics, of the entire dataset.
- **Streams**. A text stream, such as the posting or retweeting of messages on Twitter, is a text data source in which text data are continuously produced. Visually displaying such kind of text stream helps illustrate the overall trend of data over time.

Based on these types of data, visualizations are designed to assist in analysis tasks for various purposes and application domains. For example, techniques have been proposed to analyze topics, discourse, events, and sentiments, which can be further grouped into the following four categories:

- **Showing Similarity**. Techniques in this category are developed to illustrate content-wised similarities of different documents. Various similarity measurements have been proposed based on two major types of techniques, which are projection-based and semantic-oriented.
- **Showing Content**. Most text visualization techniques have been proposed to illustrate different aspects of the content of text data, such as summarizing the content of a single document and showing the topics of a corpus.
- **Showing Opinions and Emotions**. This category includes techniques that summarizing the sentiment or emotional profiles of persons based on the text data they produced.
- **Exploring the Corpus**. Many text data exploration systems have been developed to help analysts or end users to efficiently explore text data. Many of these techniques are based on information retrieval techniques and enable a visual query approach to retrieve information based on user interests.

1.3 Book Outline

This book presents a systematic review of existing text visualization techniques, from the elementary to the profound, and covers most of the critical aspects of visually representing unstructured text data.

The book consists of seven chapters in total. In particular, we first provide an overview of the entire text visualization field to our readers in Chap. 2 by introducing a taxonomy based on existing techniques and briefly survey and introduce the typical techniques in each category. Transforming the unstructured text data into a structured form is a typical approach and the first step for visualizing data. In Chap. 3, we introduce typical data structures and models used for organizing the text data and the corresponding approaches of data transformation. Starting in Chap. 4, we introduce the detailed text visualization techniques following the taxonomy introduced in Chap. 2. In particular, we introduce the techniques for visualizing document similarity in Chap. 4, for showing document content in Chap. 5, and for visualizing sentiments and emotions in Chap. 6. Finally, we conclude the book in Chap. 7.

We suggest that readers finish the first three chapters one by one before reading the rest of the book. This suggestion will help the readers obtain basic ideas on text visualization as well as the current techniques and research trends in the field. Thereafter, readers could choose to read Chaps. 4–7 based on their interests.

References

1. Averbuch, M., Cruz, I., Lucas, W., Radzyminski, M.: As You Like It: Tailorable Information Visualization. Tufts University, Medford (2004)
2. Card, S., Mackinlay, J., Shneiderman, B.: Readings in Information Visualization: Using Vision to Think. Morgan Kaufmann, Los Altos (1999)
3. Chen, C.: Top 10 unsolved information visualization problems. IEEE Comput. Graph. Appl. 12–16 (2005)
4. Ellis, G., Dix, A.: A taxonomy of clutter reduction for information visualisation. IEEE Trans. Vis. Comput. Graph. **13**(6), 1216–1223 (2007)
5. Hand, D., Mannila, H., Smyth, P.: Principles of Data Mining. The MIT Press, Cambridge (2001)
6. Hersh, W.: Information Retrieval: A Health and Biomedical Perspective. Springer, Berlin (2009)
7. Keim, D., Mansmann, F., Schneidewind, J., Ziegler, H.: Challenges in visual data analysis. In: Tenth International Conference on Information Visualization, 2006. IV 2006, pp. 9–16. IEEE (2006)
8. Koffka, K.: Principles of Gestalt Psychology. Psychology Press, Milton Park (1999)
9. Lee, B., Czerwinski, M., Robertson, G., Bederson, B.B.: Understanding research trends in conferences using paperlens. In: CHI'05 Extended Abstracts on Human Factors in Computing Systems, pp. 1969–1972. ACM (2005)
10. Shneiderman, B.: The eyes have it: a task by data type taxonomy for information visualizations. In: IEEE Symposium on Visual Languages, 1996. Proceedings, pp. 336–343. IEEE (1996)
11. Thomas, J., Cook, K.: Illuminating the Path: The Research and Development Agenda for Visual Analytics. IEEE Computer Society, Los Alamitos (2005)
12. Treisman, A.: Preattentive processing in vision. Comput. Vis. Graph. Image Process. **31**(2), 156–177 (1985)
13. Ware, C.: Information Visualization: Perception for Design. Morgan Kaufmann, Los Altos (2004)
14. Zhang, J.: The nature of external representations in problem solving. Cogn. Sci. **21**(2), 179–217 (1997). doi:10.1016/S0364-0213(99)80022-6. http://www.sciencedirect.com/science/article/B6W48-3Y2G07V-S/2/7f030b21efbc2de49b719126601212a5

Chapter 2
Overview of Text Visualization Techniques

Abstract The increasing availability of electronic document archives, such as web-pages, online news sites, blogs, and various publications and articles, provides an unprecedented amount of information. This situation introduces a new challenge, which is the discovery of useful knowledge from large document collections effectively without completely going through the details of each document in the collection. Information visualization techniques provide a convenient means to summarize documents in visual forms that allow users to fully understand and memorize data insights. In turn, this process facilitates data comparison and pattern recognition. Many text visualization techniques have been extensively studied and developed for different purposes since the 1990s. In this chapter, we briefly review these techniques to provide an overview of text visualization. Our survey is based on studies summarized in the online text visualization browser (http://textvis.lnu.se/). We classify different text visualization techniques regarding their design goals, which largely group existing techniques into five categories. These categories include techniques developed (1) for visualizing document similarity, (2) for revealing content, (3) for visualizing sentiments and emotions of the text, (4) for exploring document corpus, and (5) for analyzing various domain-specific rich-text corpus, such as social media data, online news, emails, poetry, and prose. Based on this taxonomy, we introduce the details of the primary text visualization research topics in the following chapters of this book.

2.1 Review Scope and Taxonomy

In this book, we have reviewed over 200 papers summarized in the Text Visualization Browser [61] (Fig. 2.1), which was developed by the ISOVIS Research Group from Linnaeus University in Sweden. This online tool provides the most comprehensive and up-to-date summary of text visualization that has been published. This browser enables a user to filter these techniques interactively according to tasks, data, application domain, and visual design styles, which greatly support the writing of this book, especially this chapter.

C. Nan and W. Cui, *Introduction to Text Visualization*, Atlantis Briefs
in Artificial Intelligence 1, DOI 10.2991/978-94-6239-186-4_2

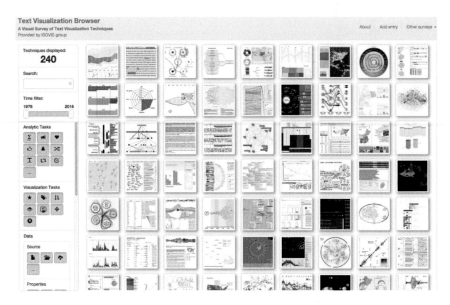

Fig. 2.1 TextVis Browser, an interactive online browser of existing text visualization techniques

Among all the studies listed in the Text Visualization Browser, a set of 120 core text visualization papers were reviewed; these were published in key conferences and journals in three different but highly related areas, namely, visualization, data mining, and human-computer interaction. In particular, our review focuses on the related papers published in (1) visualization-related conferences and journals, such as IEEE International Conference on Information Visualization, IEEE International Conference on Visual Analytics Science and Technology, IEEE Transactions on Visualization and Computer Graphics, IEEE Computer Graphics and Applications, Computer Graphics Forum, and Information Visualization; (2) data mining-related conferences and journals, such as ACM sigKDD Conference on Knowledge Discovery and Data Mining, IEEE Transactions on Knowledge and Data Engineering, SIAM International Conference on Data Mining, IEEE International Conference on Data Engineering, and ACM International Conference on Information and Knowledge Management; (3) human-computer-interaction related conferences and journals including ACM sigCHI Conference on Human Factors in Computing Systems and ACM International Conference on Intelligent User Interfaces.

Text Visualization Browser provides four different ways to categorize existing techniques, i.e., categorizing by task (either analysis or visualization), by data to be visualized, by application domain, and by style of visualization design. However, providing a clear taxonomy with minimum overlap among different technique categories is difficult. To address this issue, we provide a simple yet clear taxonomy based on development goals and purposes of the existing works. In particular, we first separate the techniques into two parts, namely, visualization or interaction technique and various systems developed for different domains by employing these techniques. In particular, in terms of visualization technique, we classify related research into three categories based on goals. These categories include techniques developed for (1) visualizing document similarity, (2) revealing the content of the text data, and (3) showing sentiments and emotions. In this way, we discuss different types of techniques in this book clearly, as well as illustrate their applications and show examples of using these techniques together for solving application problems in different domains.

The rest of this chapter and the book is presented by following this taxonomy. We first briefly review existing techniques and applications in this chapter and describe the detail of major and important techniques in the subsequent chapters.

2.2 Visualizing Document Similarity

Representing content similarities at the document level is one of the most traditional visualization techniques produced for summarizing document collections. These visualizations share a similar representation in which documents are visualized as points on a low-dimensional (2D or 3D) visualization plane. The distance between each pair of points represents the similarities between the corresponding two documents, and follows the rule of the closer, the more similar. Many similar techniques have been extensively studied and have been categorized as either (1) projection-oriented or (2) semantic-oriented.

2.2.1 Projection Oriented Techniques

Projection oriented techniques visualize documents through a dimension reduction procedure. In these techniques, a document is represented as a bag of words and formally described by an N-dimensional feature vector. To compute this vector, a set of most informative words W ($|W| = N$) that best differentiates each document (i.e., best captures the features of different documents) is ranked out from the entire document collection based on "Term Frequency Inverse Document Frequency (TF-IDF)". As a well-known numerical statistic method designed to help extractive word features for document classification [78], the process computes a TF-IDF score for each word in the document collection. The word with a higher score is considered to be more informative, i.e., more useful than other words for classifying different documents. Before computing TF-IDF, stop words are removed and word stems are extracted to ensure the informativeness of each word and the correctness of the word frequency calculation in TF-IDF. Based on these words, an N-dimensional feature vector is produced for a document with each field indicating a word with top-ranking TF-IDF score and the field value indicating its frequency in the given document. Based on this feature vector, projection-oriented techniques visualize a document from the N-dimensional feature space into a 2D or 3D visualization space via a dimension reduction algorithm.

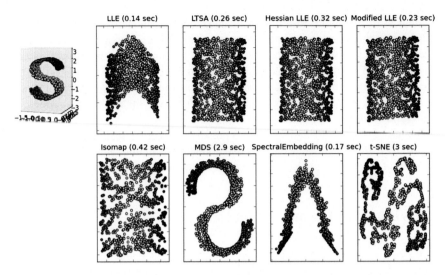

Fig. 2.2 Comparison of different Non-Linear Projection Techniques (also known as manifold learning). The results shown in this figure are is produced based on scikit-lear, which illustrates the results of projecting a "S-shaped" three dimensional dataset (*the left most figure*) onto a 2D plane. The visualization results illustrate that how the original distance in 3D space is preserved in a 2D plane via different algorithms

Generally, the projection oriented techniques can be further separated into (1) linear projections (i.e., linear dimension reduction) and (2) non-linear projections (also known as manifold learning or non-linear dimension reduction). Representative techniques in linear projection include *Principle Component Analysis (PCA)* [51] and *Linear Discriminant Analysis* [5]. Both techniques could be formulated in consistent form in which pair-wise distances between data items are maximized and guided by weights that indicate the importance of separating pairs of points in the results [58]. These techniques, although effective in terms of computation, usually fail to capture data similarities when they are non-linear. Therefore, many non-linear projection techniques have been developed and extensively studied. Existing methods could be further classified into (1) distance-oriented techniques and (2) probabilistic formulation-based techniques. A comparison of different techniques is shown in Fig. 2.2.

Specifically, the distance oriented non-linear projection techniques such as *Multidimensional Scaling (MDS)* [60] and *Locally Linear embedding (LLE)* [92] introduce different methods to preserve the distances in the high-dimensional feature space in a low (2D or 3D) dimensional visualization space. The probabilistic formulation based techniques such as *Stochastic Neighbor Embedding (SNE)* [45] and *t-Distributed Stochastic Neighbor Embedding (t-SNE)* [73] formulate document similarity via statistical models, in which the similarity between two documents i and j is captured by the conditional probability of $P_{(j|i)}$, i.e., given that document i the probability of document j are in the neighborhood of i in the feature space. Compared with distance oriented techniques, these probabilistic based approaches can be more effectively computed and can produce results with improved quality [73].

2.2.2 Semantic Oriented Techniques

These approaches represent document similarity via latent topics extracted from text data. Studies in this direction are mainly inspired by topic modeling techniques such as *Probabilistic Latent Semantic Analysis (PLSA)* [46], *Latent Dirichlet Allocation (LDA)* [6], *Spherical Topic Model (SAM)* [87], and *Non-Negative Matrix Factorization (NMF)* [64]. Although widely used for analysis, these topic modeling techniques are not designed for visualization purpose, thereby directly showing the analysis results is usually difficult for users to interpret. For example, PLSA and LDA analyze topics in a simplex space which is shown as a triangle on the 2D Euclidean visualization plane. Therefore, these methods cannot display more than three topics a time (Fig. 2.3d). Semantic oriented techniques have proposed to produce improved visual representations. These techniques were pioneered by *Probabilistic Latent Semantic Visualization (PLSV)* [47] which embeds the latent topics and docu-

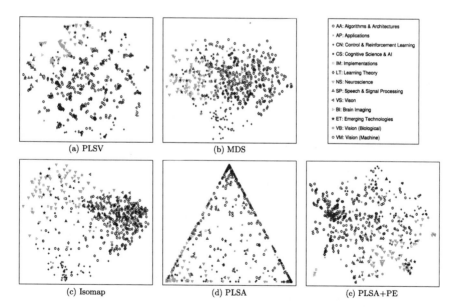

(a) PLSV   (b) MDS

○ AA: Algorithms & Architectures
▸ AP: Applications
∗ CN: Control & Reinforcement Learning
• CS: Cognitive Science & AI
▫ IM: Implementations
◇ LT: Learning Theory
▽ NS: Neuroscience
△ SP: Speech & Signal Processing
◁ VS: Vison
▹ BI: Brain Imaging
★ ET: Emerging Technologies
◦ VB: Vision (Biological)
○ VM: Vision (Machine)

(c) Isomap   (d) PLSA   (e) PLSA+PE

Fig. 2.3 Visualizing document similarity based on semantic oriented techniques (**a, d, e**) and non-linear projection (**b, c**). Each *point* in the diagram is a document colored by their primary topics that are extracted based on topic modeling. The distances between documents encode their similarities, following the rule of "the more similar, the closer". This figure represent visualizations of the same data, i.e., papers published in NIPS, which was first published in [47]

ments in the generic Euclidean space at the same time as the distances directly encode similarities among documents, regarding their shared topics (Fig. 2.3a). Following this work, other techniques such as Spherical Semantic Embedding (SSE) [63] were also developed, which provides improved approximations of similarities among documents. When compared to the projection oriented techniques (Fig. 2.3b, c), semantic based approaches usually produce more meaningful results that are easier for users to understand.

2.3   Revealing Text Content

Visually representing the content of a text document is one of the most important tasks in the filed of text visualization. Specifically, visualization techniques have been developed to show the content of documents from different aspects and at different levels of details, including summarizing a single document, showing the words and topics, detecting events, and creating storylines.

2.3.1   Summarizing a Single Document

Existing visualization techniques summarize a document through two main aspects: (1) content such as words and figures and (2) features such as average sentence length and number of verbs.

In terms of showing the content of a document, Collins et al. [19] introduce DocBurst which decompose a document into a tree via its innate structures such as sections, paragraphs, and sentences that are illustrated in a SunBurst visualization [97] (Fig. 2.4). Rusu et al. [93] visualize the content of a document via a node-link diagram based on a semantic graph extracted from the document. Strobelt et al. [99] introduce a system that transforms a document into cards, in which the content of the document is summarized via keywords and critical figures that are extracted from the document (Fig. 2.5). Stoffel et al. [98] propose a technique for producing the thumbnail of a document based on keyword distortion. This technique produces a focus+context representation of document at the page level. In particular, on each page of the document, important words are shown in a larger font whereas the rest ones are suppressed as the context, thereby compressing the entire page into a small thumbnail without losing the key information of each document page.

Despite the aforementioned studies, document fingerprint [50, 55, 80] is another typical visualization technique that is developed to summarize a single document. Instead of showing words and figures, this technique captures the key features of a document from multiple aspects through a heatmap visualization in which each cell represents a text block (e.g., a paragraph or a sentence) with color showing its feature value (Fig. 2.6). A set of linguistic features were used to measure the document from different aspects, as summarized in [55], including (1) statistical

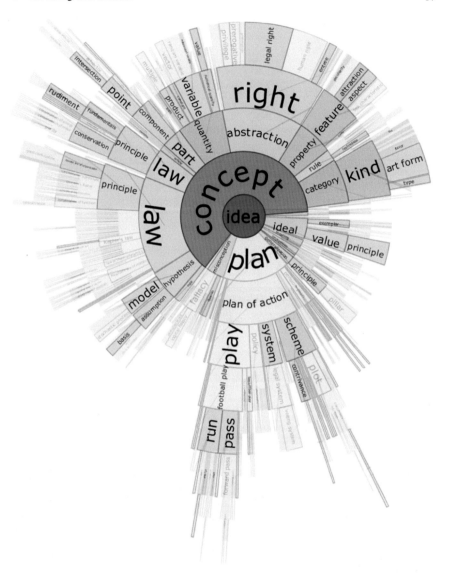

Fig. 2.4 DocuBurst visualization of a science textbook rooted at idea. A search query for words starting with pl has been performed. Nodes matching the query are highlighted in *gold*

features such as average word length, average number of syllables per word and average sentence length, (2) vocabulary features such as the frequencies of specific words and vocabulary richness measured by Simpson Index, and (3) syntax features computed based on syntax trees [54].

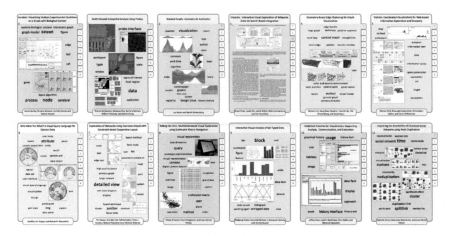

Fig. 2.5 Summarization of the IEEE InfoVis 2008 proceedings corpus in Document Cards (a portion). Referring to [98] for the complete visual summarization of the whole proceeding

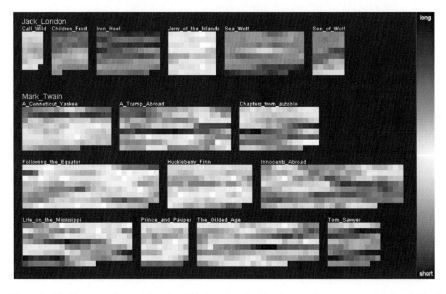

Fig. 2.6 The literature fingerprint visualization showing the feature "average sentence length" of books written by Jack London and Mark Twain

2.3.2   Showing Content at the Word Level

Directly illustrating the keywords of a document is the most intuitive approach to present document content. Existing visualization techniques in this category are largely developed to address three general problems: (1) how to represent the words esthetically in a visual form to clearly depict the content of the text; (2) how to

Fig. 2.7 Wordle visualization of a bag of words extracted from text data

summarize and represent the semantic relationships such as "A is B" and "A of B" between words in the text, and (3) how to reveal word-level patterns such as repetitions and co-occurrences.

TagCloud [53] is one of the most intuitive and commonly used techniques for visualizing words. It illustrates a bag of words that summarize the content of the input text data in a cloud form, in which words, with font size indicating their importance, are packed together without any overlap. Traditional TagCloud aligns words line by line, which is most commonly used in webpages to show the content of, for example, the current web. Different packing strategies will result in various types of TagClouds [13, 14, 20, 35, 104, 114], among which Wordle is the state-of-the-art technique that produces aesthetic word packing results by precisely calculating the word boundary and randomly inserting the word into empty spaces guiding by a spiral line (Fig. 2.7). Despite these static techniques, dynamic word clouds [24] are also developed for showing the changes of the text content of a streaming corpus such as Twitter and publication dataset over time.

Although widely used, TagClouds fail to uncover the word relationship as usually the words are randomly placed. Therefore, many tree or graph based visualization techniques are introduced. For example, WordTree [109] (Fig. 2.8) summarizes text data via a syntax tree in which sentences are aggregated by their sharing words and split into branches at a place where the corresponding words in the sentences are divergent. Another example is the PhraseNet [103] (Fig. 2.9). This visualization employs a node-link diagram, in which graph nodes are keywords and links represent relationships among keywords which are determined by a regular expression indicated by users. For example, as shown in Fig. 2.9, a user can select a predefined regular expression from a list to extract a relationship such as "X and / is / of Y" or by inputting the regular expression by their own.

Despite relationships, visualizations are also developed to illustrate highly detailed patterns such as word co-occurrences and repetitions [4, 49, 108]. For example, Wattenberg introduced the Arc diagram (Fig. 2.10), which is one of the earliest visu-

Fig. 2.8   Word tree of the King James Bible showing all occurrences of *love the*

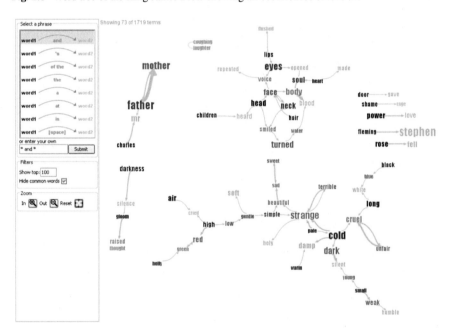

Fig. 2.9   The Phrase Net user interface applied to James Joyce Portrait of the Artist as a Young Man. The user can select a predefined pattern from the list of patterns on the left or define a custom pattern in the box below. This list of patterns simultaneously serves as a legend, a list of presets and an interactive training mechanism for regular expressions. Here the user has selected X and Y, revealing two main clusters, one almost exclusively consisting of adjectives, the other of verbs and nouns. The highlighted clusters of terms have been aggregated by our edge compression algorithm [103]

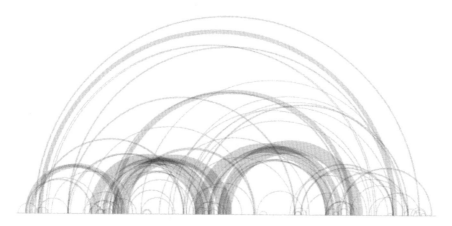

Fig. 2.10 The arc diagram visualization of a HTML webpage in which repeated strings (e.g., words or sentences) are connected by arcs

alizations designed to illustrate repetition patterns. It represents an input string in a raw and connects the repeated symbols or words via arcs, thereby illustrating a clear visual pattern when repetition occurs.

2.3.3 Visualizing Topics

Accompanying the rapid development of topic analysis techniques, visualizing topics has become a highly interesting research direction in the field of text visualization in recent years. Compared with word-level visualizations, showing topics helps to capture more semantics of the data, thereby producing text visualizations that are easier to interpret. A growing number of visualization techniques are developed to (1) summarize and explore static topic information, (2) illustrate the topic dynamics over time, (3) help with topic comparison, and (4) illustrate events and storylines.

Even before the invention of modern topic modeling techniques such as PLSA [46] and LDA [6], visualization systems such as Topic Island™ [76] and IN-SPIRE™[1] had been introduced to illustrate and explore static topic themes extracted from text data. Research in this direction is significantly accelerated by the development of topic analysis techniques. Many attempts have been made to represent the topic analysis results. In particular, Cao et al. introduced ContexTour [67] (Fig. 2.11), FacetAtals [17], and SolarMap [15] in a row based on a so-called "multifaceted entity-relational data model". In particular, they decompose the text corpus into this data model based on a series of text analysis approached including (1) topic analysis, (2) name entity identification, and (3) word co-occurrence detection. The resulting visualization illustrates static topics and their corresponding relationships from

---

[1]http://in-spire.pnnl.gov/.

Fig. 2.11 Visualizing research topics in a publication dataset of papers published in the computer science conferences and journals in 2005 in ContexTour, in which the background contour produces a density map based on kernel density estimation, showing the underlying distribution of the words. Topics are shown as TagCoulds on *top* of the contour map

multiple information facets. Following these works, many similar techniques such as VisClustering [65] and TopicPanoram [69] were also developed. When compared to the aforementioned techniques, these works employed similar visual designs and provided similar functions in terms of topic representation and exploration, but were developed to focus on different analysis tasks (Fig. 2.12).

Capturing the topic dynamics is another research direction that attracts great attention in the field of text visualization. In particular, ThemeRiever [42] is one of the earliest techniques developed to show how the frequency of the keywords are changed over time. It visualizes a set of keywords as stripes (i.e., themes) whose thicknesses change over time, indicating the change of frequencies of the corresponding keywords, which are shown on top of the stripe. This design was later extended by Liu

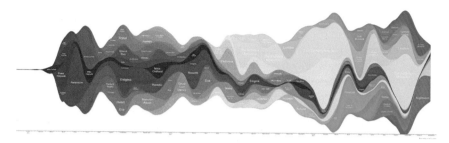

Fig. 2.12 ThemeRiver [42]: keywords in a document collection are shown as *colored "stripes"*, i.e. themes, with width indicating the occurrence frequency of keywords at different times

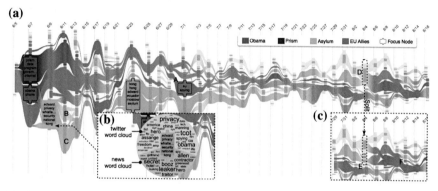

Fig. 2.13 RoseRiver, a visual analytics system for exploring evolutionary hierarchical topics. **a** Overview of the Prism scandal (June 5–Aug 16, 2013). Four *colors* represent the four major topics. Topics are displayed as *vertical bars*. The *color stripes* represent the evolving relationships between topics. **b** Comparison of the prominent keywords in tweets and news articles of the topic. The arc lengths encode the news article and tweet numbers (in log scale). **c** The new layout generated by splitting the gray topic

et al. [71] by using strips to represent changing topics, on top of which topic keywords are visualized as TagClouds. Following this research direction, Cui et al. [22] introduce TextFlow, the state-of-the-art design that captures the dynamics of topic evolution patterns such as splitting and merging over time. Subsequently, the authors refined their original design and algorithm to produce a hierarchical topic flow [23] that supports different levels of detail. This technique precisely captures the topic evolution patterns as shown in Fig. 2.13. Despite showing the overview of topic evolution trend, visualization techniques were also introduced to capture other temporal patterns in text data. For example, Gad et al. [34] introduced ThemeDelta, which integrates with topic modeling algorithms to identify change points when significant shifts in topics occurred. Liu et al. [68] introduced a technique for exploring topic lead-lag, i.e. a pattern illustrating how a former topic result in the occurrence of a latter topic

When multiple document collections are visualized together at the same time, a spontaneous analysis task is to compare to find their common and distinct topics. To this end, many visualization techniques are introduced. Diakopoulos et al. [25] develop Compare Clouds, which is a TagCloud visualization designed to compare the topic keywords of two sets of input documents. Oelke et al. [81] also introduce a visual analysis system for comparing and distinguishing different document collections. In particular, it detects discriminative and common topics and visualizes each topic in a circular glyph, in which topic keywords are shown as a TagCloud. Glyphs are laid out based on topic similarities, i.e., the glyphs of similar topics are placed close to each other whereas the dissimilar ones are separated apart. In this way, common topics are placed at the center of the view and the discriminative ones are clearly separated into different topic regions, thereby forming a visualization that facilitates topic differentiations and comparisons.

2.3.4 Showing Events and Storyline

Finding topics in a collection of documents is generally a clustering approach. Documents that have similar contents (essentially using similar words) are clustered together to constitute a topic or a theme. By contrast, event analysis focuses on a different type of information that has time and space as the primary attributes [48]. An event is generally considered as an occurrence at a given space-time that is perceived by an observer to have a beginning and an end. Human beings are gifted with the ability to perceive and make sense of real-world activities as consisting of discrete events with orderly relations [116]. Thus, when visualizing events, researchers more focus on understanding the four Ws that characterize these events: who, what, when, and where.

A large amount of structured or semi-structured textual data explicitly contain event information, such as crime incidents or accidents recorded by police departments [11, 66], patient records [21, 32, 39, 86, 111–113], and customer purchase logs [12]. For these datasets, the primary goal of visualization is to provide visual summaries and to support efficient queries. For example, as an early work, Life-Lines [86] provides a general visualization that allows users to explore details of a patient's clinical records (Fig. 2.14). A subsequent version, LifeLines2 [106] enhances the visualization and introduces three general operators, namely, align, rank, and filter, for interactive exploration of categorical, health-related datasets. PatternFinder [32] is designed to help users visually formulate queries and find temporal event patterns in medical record datasets. LifeFlow [112] and Outflow [111] aggregate multiple event sequences into tree or graph visualizations and provide users with highly scalable overviews (Fig. 2.15).

In real-world situations, people also often segment activities into events at multiple timescales [62]. Small events are grouped to constitute a large and complex event, such as acts in a play. In addition, events may also share elements, such as participants

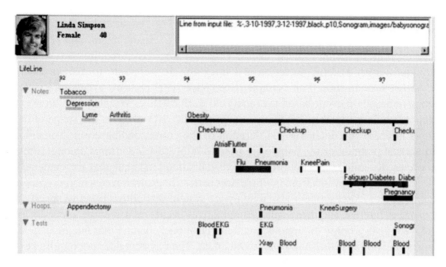

Fig. 2.14 Screenshot of LifeLines [86]: *colored horizontal bars* show the time of occurrence and duration of clinical events for a patient, such as medical incidents, treatments, and rehabilitation. Additional information is encoded by the height and color of individual bars. Multiple facets of the records, such as notes and tests, are stacked vertically, and can be expanded and collapsed as needed

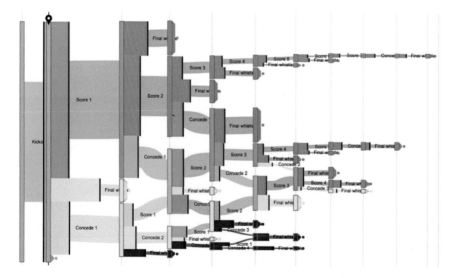

Fig. 2.15 Screenshot of Outflow [111] visualization that shows the scores of Manchester United in the 2010–2011 season. *Green* indicates pathways of winning, while *red* shows pathways of losing

and locations, with one another. To represent relationships between events, Burch et al. [12] use a horizontally oriented tree layout to represent the hierarchical relationships of transaction sequences along the timeline. André et al. [2] present Continuum to visually organize large amounts of hierarchical events and their relationships.

Recently, significant research has been conducted to extract and explore events in unstructured textual data, such as news articles [28, 95] and microblogs [26, 75]. Since event information in these data is generally implicit, it requires text mining techniques, such as topic detection and tracking, are necessary to extract events for further visualization. For example, EventRiver [72] presents events in a river-metaphor based on event-based text analysis. In EventRiver visualization, an event is represented by a bubble that floats on a horizontal river of time. The shape of the bubble indicates the intensity and duration of the corresponding event. The color and vertical position of the bubble are used to indicate relationships among different events. Krstajic et al. [59] propose a visualization technique that incrementally detects clusters and events from multiple time series. To explore events in microblogs, Marcus et al. [75] describe a visual analytics system that allows users to specify a keyword of interest, and then visually summarizes events related to the query. DÖrk et al. [26] also propose an approach called visual backchannel that integrates text, images, and authors extracted from Twitter data. Their system does not require keywords of interest. Instead, it monitors evolving online discussions on major events, such as political speeches, natural disasters, and sport games. Twitter posts are summarized as a temporally adjusted stacked graph. Related authors and images are also visualized as a People Spiral and an Image Cloud, respectively, to help users track the event evolutions. LeadLine (Fig. 2.16) combines topic analysis and event detection techniques to extract events from social media data streams. Various information, such as topic, person, location, and time, is identified to help users reason about the topic evolutions.

Recently, storyline visualization has emerged and attracted significant attention. Unlike events and topics that focus on activities and themes, storyline visualizations switch the focus to entities and relationships between entities. In a typical storyline visualization, x-axis represents time, and an entity is represented as a line that extends

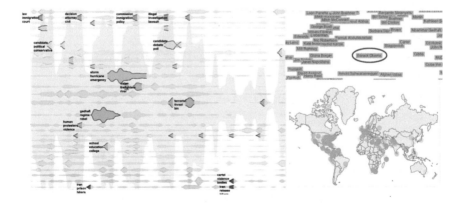

Fig. 2.16 Screenshot of LeadLine visualization that summarizes CNN news stories from Aug 15, 2011 to Nov 5, 2011. *Bottom right* locations that are related to President Obama are marked on the map. *Left* events that are related to the president are highlighted by *color-coded bursts*

horizontally from left to right. The relationships between entities, which may change over time, are encoded by the vertical distances between the lines. Figure 1.1 shows an example of such visualizations. The figure shows the main story in the book of *Lord of the Rings*. Although this chart is manually made by the author, it inspires a set of approaches that aim to automatically generate similar visualizations. For example, Ogievetsky [83] builds an online tool to allow users to interactively generate and adjust a storyline visualization. However, to find a visually satisfying arrangement of lines is the key issue for this visualization. To solve this problem, Ogawa and Ma [82] raise several criteria and propose a rule-based algorithm to automatically generated a storyline-like visualization to help experts track software evolution (Fig. 2.18). Later, Tanahashi and Ma [101] further formulate the storyline layout process as an optimization problem and solve the problem with a genetic method. Although time-consuming, their algorithm can generate results that are comparable to those handmade by XKCD. Based on their work, another optimization process is proposed by Liu et al. [70]; This process is time efficient enough to support real-time interaction and still be able to maintain the same level of aesthetics (Fig. 2.19). Recently, Tanahashi et al. [100] have extended storyline visualizations to streaming data, and further provide users with the ability to follow and reason dynamic data.

Fig. 2.17 XKCD's movie narrative chart of *Lord of the Rings* [77]

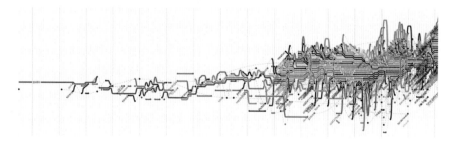

Fig. 2.18 Storyline that shows the development of Python source codes [82]

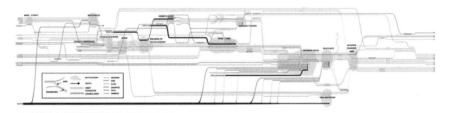

Fig. 2.19 Reproduction of the same chat in Fig. 2.17 using StoryFlow algorithm [82]

2.4   Visualizing Sentiments and Emotions

Many visualization techniques have been developed to illustrate the change of sentiments over time regarding to a given streaming text corpus such as news corpus, review comments, and Twitter streams. This goal can be achieved, as shown in Fig. 2.20, by showing the sentiment dynamics in a time-series diagram, in which the time-series curve illustrates the change of sentiment scores computed across the entire dataset at different time points. However this simple visualization is too abstract to display information details such as the causes behind the sentiment shifts. Therefore, many other more advanced techniques have been introduced to illustrate and interpret the sentiment dynamics from different perspectives.

Most techniques are developed to compute and visualize the sentiments of a focal group of people based on the text data produced by them. The resulting visualization forms a "happiness indicator" that captures the sentiment change of the focal group over time. For example, Brew et al. [9] introduce SentireCrowds, which represents the sentiment changes of a group of twitter users from the same city in a timeline view

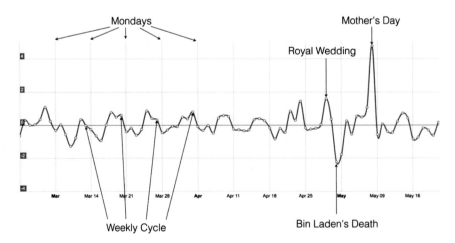

Fig. 2.20 Sentiment Indexing of Twitter data in a time-series diagram. This figure shows that the public sentiment may change dramatically regarding to different events in our real-life

and summarizes the potential underlying event that causes the changes in a multi-level TagCloud designed based on Treemap. Zhao et al. [118] introduce PEARL, which visualizes the change of a person's emotion or mood profile derived from his tweets in a compound belt visualization. The belt groups a set of emotion bands, each indicating a type of emotion differentiated by colors. The thickness of the band changes over time, representing the portion of the corresponding emotion at different time. Guzman et al. [40] visualize the change of emotions of groups of different developers in various software development projects. Hao et al. [41] analyze sentiments via geo-temporal term associations based on a streaming dataset of customer's feedback. Kempter et al. [57] introduce a fine-grained, multi-category emotion model to classify users' emotional reactions to public events overtime and to visualize the results in a radar diagram, called EmotionWatch, as shown in Fig. 2.22.

Despite the preceding visualizations, some visual analysis systems have also been developed to assist with dynamic sentiment analysis. For example, Wanner et al. [107] develop a small multiple visualization view to conduct a semi-automatic sentiment analysis of large news feeds. In this work, a case study on news regarding the US presidential election in 2008 shows how visualization techniques will help analysts draw meaningful conclusions without exerting effort to read the content of the news. Brooks et al. [10] introduce Agave, a collaborative visual analysis system for exploring events and sentiment over time in large Twitter datasets. The system employs multiple co-ordinated views in which a streamgraph (Fig. 2.21) is used to summarize the changes of the sentiments of a subset of tweets queried based on users' preferences. Zhang et al. [117] introduce a spacial-temporal view for visualizing the sentiment scores of micro-blog data based on an electron cloud model intruded in physics. The resulting visualization maps a single sentiment score to a position inside a circular visualization display.

More sophisticated systems are also developed to analyze the change of sentiments based on streaming text data. For example, Rohrdantz et al. [91] introduce a visual analysis system to help users to detect interesting portions of text streams,

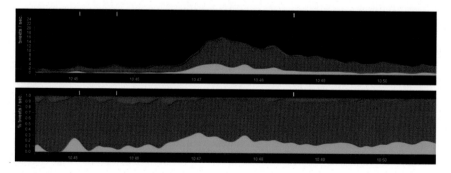

Fig. 2.21 Sentiment streamgraphs for the keyword search Flacco, the Super Bowl MVP in a Twitter dataset using Agave [10]. Negative is *red*, neutral is *gray*, and positive is *blue*. *Top* overall frequency of tweets, divided by sentiment type. *Bottom* sentiment as percent of overall volume

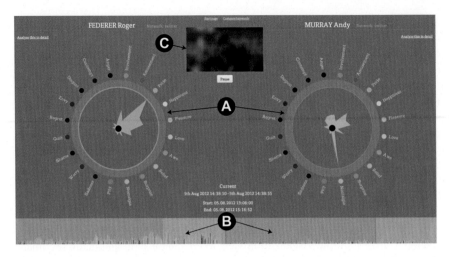

Fig. 2.22 Comparison of two emotion profiles of Roger Federer and Andy Murray (two tennis athletes) after a tennis game in EmotionWatch [57]; (*A*) the EmotionWatches, (*B*) timelines showing the two emotion flows, and (*C*) video

regarding to the change of sentiments, data density, and context coherence based on a set of features extracted from the text. Wang et al. [105] introduce SentiView, which employs advanced sentiment analysis techniques as well as visualization designs to analyze the change of public sentiments regarding popular topics on the Internet. Other systems are designed for analyzing sentiment divergences (i.e., conflicting of opinions) that occur between two groups of people. For example, Chen et al. [18] introduce the first work in this topic based on a simple time-series design that summarizes the overall conflicting opinions based on the Amazon review data. Following this topic, Cao et al. [16] introduce a more advanced technique called SocialHelix, which extracts two groups of people having the most significant sentiment divergence over time from Twitter data and illustrates their divergence in a Helix visualization to show how the divergence occurred, evolved, and terminated.

In terms of application, a large set of such techniques are developed to represent the customer's sentiments based on the review data. Alper et al. [1] introduce OpinionBlocks, an interactive visualization tool to improve people's understanding of customer reviews. The visualization progressively discloses text information at different granularities from the keywords to the phrases in which the keywords are used, and to the reviews containing the phrases. The information is displayed within two horizontal regions, representing two types (positive and negative) of different sentiments. Gamon et al. [36] introduce Pulse for mining topics and sentiment orientation jointly from free text customer feedback. This system enables the exploration of large quantities of customer review data and was used for visually analyzing a database for car reviews. Through this system, users can examine customer opinion at a glance or explore the data at a finer level of detail. Oelke et al. [79] analyze to determine customers' positive and negative opinions through the comments or rat-

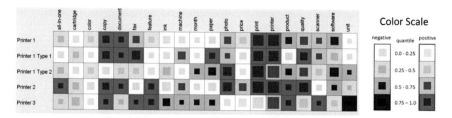

Fig. 2.23 Summary Report of printers: each *row* shows the attribute performances of a specific printer. *Blue color* represents comparatively positive user opinions and *red color* comparatively negative ones (see *color scale*). The size of an *inner rectangle* indicates the amount of customers that commented on an attribute. The larger the rectangle the more comments have been provided by the customers

ings posted by the customers. This system visualize the analysis results in a heatmap view showing both volume of comments and the summarized sentiments (Fig. 2.23). Wu et al. [115] introduce OpinionSeer, which employs subjective logic [52] to analyze customer opinions on hotel rooms based on their review data inside a simplex space, which is visualized in a triangle surrounded by context about the customers such as their ages and their countries of origin. More generic systems are also developed. For example, Wensel [110] introduce VIBES, which extracts the important topics from a blog, and measures the emotions associated with those topics that are illustrated through a range of different visualization views. Makki et al. [74] introduce an interactive visualization to engage the user in the process of polarity assignment to improve the quality of the generated lexicon used for sentiment or emotion analysis via minimal user effort.

2.5   Document Exploration Techniques

With a large document collection, how to effectively explore the data to find useful information or insightful data patterns is always a challenge that attracts many research attentions. Many visualization systems are designed to support an effective exploration of big text corpus. Many studies are focused on inventing or improving the text data exploration techniques. A large category of them are query-based systems in which full text indices are built so that users can query to retrieve data based on their interests. Based on these techniques, many systems are developed to explore text collected from various application domains. In this section, we review these exploration techniques as well as their applications.

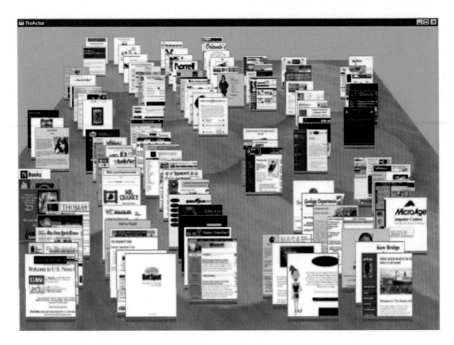

Fig. 2.24 Data Mountain visualization shows 100 webpages

2.5.1 Distortion Based Approaches

As early as the 1990s, some distortion based techniques have been developed to assist in text data exploration. For example, Robertson and Mackinlay introduced Document Lens [90] which introduced a focus+context design inspired by the magnifier lens. In this visualization, the focused content of a document is shown in details in the view center, surrounded by other parts of the content that provides an overall impression of the text data. Despite showing the content, a similar idea is also used to visualize documents. For example, Data Mountain [89] employs a focus+context view based on a prospective projection, in which the focused documents are shown in a larger size with additional details in front of other documents, whereas the unfocused ones are shown at the back side in a smaller size (Fig. 2.24). Users can interactively switch between the focus and context by clicking on the documents.

2.5.2 Exploration Based on Document Similarity

Exploring documents in a similarity view is another early but popular technique that is extensively used in many text visualizations. These systems, such as InfoSky [3] and $IN\text{-}SPIRE^{TM}$ [44] and ForceSPIRE [31], use an overview that summarizes the entire document collection based on document similarities (see, Sect. 2.1) and employs a multiple coordinated view to reveal the document details from various aspects such as keywords and topics. Users can navigate through the similarity view based on interactions such as zooming and panning, thereby showing different levels of details [30].

2.5.3 Hierarchical Document Exploration

Exploring big document collection based on hierarchical clustering is another commonly used approach. For example, Paulovich and Minghim introduce HiPP [85], which lays out documents via a hierarchical circle packing algorithm, in which a document is shown as a circle. Dou et al. [29], Brehmer et al. [8], as well as Pascual-Cid and Kaltenbrunner [84] develop different types of document exploration systems that are all based on hierarchical clustering. In these systems, documents are hierarchically clustered based on their topic similarity and the cluster results are shown in a tree view to guide the data navigation.

2.5.4 Search and Query Based Approaches

Full text search and document query are also widely used to support document exploration since the very beginning of the text visualization [43, 94, 96]. Instead of showing a ranked list of related documents regarding the query keywords, most of the existing visualization techniques transform the search results into a visual representation to illustrate the insight of content relationships among documents. Graph layout and projection-based approach are commonly used to represent the search and query results showing the relationships among documents [7], or text snippets such as words [88] or collections of topic keywords [33, 37]. More advanced techniques are also developed. For example, Isaacs introduced Footprints to support an effective and interactive procedure to retrieve information in a desired subject from a large document collection. In this system, the topics queried by a user are visually summarized in an iconic representation as shown in Fig. 2.25. The user can click the topic to load a collection of related document in the document list and the content of a selected document can be further shown in the document viewer. A set of filters also helps users extract the most interesting data.

Exploration in Coordinated Views. The aforementioned document exploration techniques are usually combined with multiple coordinated visualization views that illustrate different aspects of the input document collection. Thew views are usually connected together through interactions such as "Linking and Brushing".[2] In these

---

[2]"The idea of linking and brushing is to combine different visualization methods to overcome the shortcomings of single techniques. Interactive changes made in one visualization are automatically

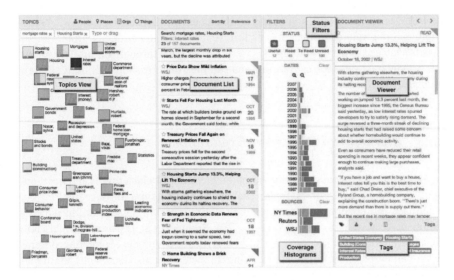

Fig. 2.25 Footprints, a topics-based document search tool, which supports exploratory search to help analysts **a** retrieve unknown information the goal of query is missing, **b** track the query results from multiple aspects to avoid missing important information and help to determine when should the query procedure be stopped

techniques, topic analysis is one of the most important aspects. For example, Dou et al. [27] introduce ParallelTopics, which guides the document exploration via multiple views. Specifically, it employs a parallel coordinates to illustrate how a document distributed in different topics, employs a TagCloud view to illustrate the keywords inside each topic, and employs the theme river to show the change of topics over time. Many other similar systems are available, such as Jigsaw [38] and IVEA [102], which are discussed in details in Chap. 4.

2.6   Summary of the Chapter

In this chapter, we reviewed more than 200 papers in the field of text visualization to provide an overview. This chapter provides readers a brief idea of what is text visualization and what is the research focus of this field. In particular, we summarize the existing works into three major categories based on the type of information to be shown in a visualization. Specifically, these techniques include those for (1) visualizing document similarities, (2) revealing text content, and (3) visualizing

---

(Footnote 2 continued)
reflected in the other visualizations. Note that connecting multiple visualizations through interactive linking and brushing provides more information than considering the component visualizations independently." [56].

semantics and emotions. We also reviewed the most commonly used text exploration techniques, including (1) distortion-based approaches, (2) exploration based on document similarity, (3) hierarchical document exploration, (4) search and query-based approaches, and (5) exploration in coordinated views. In the following chapters, we focus on detailed techniques.

References

1. Alper, B., Yang, H., Haber, E., Kandogan, E.: Opinionblocks: visualizing consumer reviews. In: IEEE VisWeek 2011 Workshop on Interactive Visual Text Analytics for Decision Making (2011)
2. André, P., Wilson, M.L., Russell, A., Smith, D.A., Owens, A., et al.: Continuum: designing timelines for hierarchies, relationships and scale. In: Proceedings of the 20th Annual ACM Symposium on User interface Software and Technology, pp. 101–110. ACM (2007)
3. Andrews, K., Kienreich, W., Sabol, V., Becker, J., Droschl, G., Kappe, F., Granitzer, M., Auer, P., Tochtermann, K.: The infosky visual explorer: exploiting hierarchical structure and document similarities. Inf. Vis. 1(3–4), 166–181 (2002)
4. Angus, D., Smith, A., Wiles, J.: Conceptual recurrence plots: revealing patterns in human discourse. IEEE Trans. Vis. Comput. Graph. 18(6), 988–997 (2012)
5. Balakrishnama, S., Ganapathiraju, A.: Linear Discriminant Analysis—A Brief Tutorial, vol. 18. Institute for Signal and information Processing, Starkville (1998)
6. Blei, D.M., Ng, A.Y., Jordan, M.I.: Latent Dirichlet allocation. J. Mach. Learn. Res. 3, 993–1022 (2003)
7. Bradel, L., North, C., House, L.: Multi-model semantic interaction for text analytics. In: 2014 IEEE Conference on Visual Analytics Science and Technology (VAST), pp. 163–172. IEEE (2014)
8. Brehmer, M., Ingram, S., Stray, J., Munzner, T.: Overview: the design, adoption, and analysis of a visual document mining tool for investigative journalists. IEEE Trans. Vis. Comput. Graph. 20(12), 2271–2280 (2014)
9. Brew, A., Greene, D., Archambault, D., Cunningham, P.: Deriving insights from national happiness indices. In: 2011 IEEE 11th International Conference on Data Mining Workshops (ICDMW), pp. 53–60. IEEE (2011)
10. Brooks, M., Robinson, J.J., Torkildson, M.K., Aragon, C.R., et al.: Collaborative visual analysis of sentiment in Twitter events. In: Cooperative Design, Visualization, and Engineering, pp. 1–8. Springer, Berlin (2014)
11. Buetow, T., Chaboya, L., OToole, C., Cushna, T., Daspit, D., Petersen, T., Atabakhsh, H., Chen, H.: A spatio temporal visualizer for law enforcement. In: Intelligence and Security Informatics, pp. 181–194. Springer, Berlin (2003)
12. Burch, M., Beck, F., Diehl, S.: Timeline trees: visualizing sequences of transactions in information hierarchies. In: Proceedings of the Working Conference on Advanced Visual Interfaces, pp. 75–82. ACM (2008)
13. Burch, M., Lohmann, S., Beck, F., Rodriguez, N., Di Silvestro, L., Weiskopf, D.: Radcloud: visualizing multiple texts with merged word clouds. In: 2014 18th International Conference on Information Visualisation (IV), pp. 108–113. IEEE (2014)
14. Burch, M., Lohmann, S., Pompe, D., Weiskopf, D.: Prefix tag clouds. In: 2013 17th International Conference on Information Visualisation (IV), pp. 45–50. IEEE (2013)
15. Cao, N., Gotz, D., Sun, J., Lin, Y.R., Qu, H.: Solarmap: multifaceted visual analytics for topic exploration. In: IEEE International Conference on Data Mining, pp. 101–110. IEEE (2011)
16. Cao, N., Lu, L., Lin, Y.R., Wang, F., Wen, Z.: Socialhelix: visual analysis of sentiment divergence in social media. J. Vis. 18(2), 221–235 (2015)

17. Cao, N., Sun, J., Lin, Y.R., Gotz, D., Liu, S., Qu, H.: Facetatlas: multifaceted visualization
    for rich text corpora. IEEE Trans. Vis. Comput Graph. **16**(6), 1172–1181 (2010)
18. Chen, C., Ibekwe-SanJuan, F., SanJuan, E., Weaver, C.: Visual analysis of conflicting opinions.
    In: 2006 IEEE Symposium on Visual Analytics Science and Technology, pp. 59–66. IEEE
    (2006)
19. Collins, C., Carpendale, S., Penn, G.: Docuburst: visualizing document content using language
    structure. Comput. Graph. Forum **28**(3), 1039–1046 (2009)
20. Collins, C., Viegas, F.B., Wattenberg, M.: Parallel tag clouds to explore and analyze faceted
    text corpora. In: IEEE Symposium on Visual Analytics Science and Technology, 2009. VAST
    2009, pp. 91–98. IEEE (2009)
21. Cousins, S.B., Kahn, M.G.: The visual display of temporal information. Artifi. Intell. Med.
    3(6), 341–357 (1991)
22. Cui, W., Liu, S., Tan, L., Shi, C., Song, Y., Gao, Z.J., Qu, H., Tong, X.: Textflow: towards
    better understanding of evolving topics in text. IEEE Trans. Vis. Comput. Graph. **17**(12),
    2412–2421 (2011)
23. Cui, W., Liu, S., Wu, Z., Wei, H.: How hierarchical topics evolve in large text corpora. IEEE
    Trans. Vis. Comput. Graph. **20**(12), 2281–2290 (2014)
24. Cui, W., Wu, Y., Liu, S., Wei, F., Zhou, M.X., Qu, H.: Context preserving dynamic word cloud
    visualization. In: IEEE Symposium on Pacific Visualization, pp. 121–128 (2010)
25. Diakopoulos, N., Elgesem, D., Salway, A., Zhang, A., Hofland, K.: Compare clouds: visual-
    izing text corpora to compare media frames. In: Proceedings of IUI Workshop on Visual Text
    Analytics (2015)
26. Dörk, M., Gruen, D., Williamson, C., Carpendale, S.: A visual backchannel for large-scale
    events. IEEE Trans. Vis. Comput. Graph. **16**(6), 1129–1138 (2010)
27. Dou, W., Wang, X., Chang, R., Ribarsky, W.: Paralleltopics: a probabilistic approach to explor-
    ing document collections. In: 2011 IEEE Conference on Visual Analytics Science and Tech-
    nology (VAST), pp. 231–240. IEEE (2011)
28. Dou, W., Wang, X., Skau, D., Ribarsky, W., Zhou, M.X.: Leadline: interactive visual analysis
    of text data through event identification and exploration. In: 2012 IEEE Conference on Visual
    Analytics Science and Technology (VAST), pp. 93–102. IEEE (2012)
29. Dou, W., Yu, L., Wang, X., Ma, Z., Ribarsky, W.: Hierarchicaltopics: visually exploring large
    text collections using topic hierarchies. IEEE Trans. Vis. Comput. Graph. **19**(12), 2002–2011
    (2013)
30. Endert, A., Burtner, R., Cramer, N., Perko, R., Hampton, S., Cook, K.: Typograph: multiscale
    spatial exploration of text documents. In: 2013 IEEE International Conference on Big Data,
    pp. 17–24. IEEE (2013)
31. Endert, A., Fiaux, P., North, C.: Semantic interaction for visual text analytics. In: Proceedings
    of the SIGCHI Conference on Human Factors in Computing Systems, pp. 473–482. ACM
    (2012)
32. Fails, J.A., Karlson, A., Shahamat, L., Shneiderman, B.: A visual interface for multivariate
    temporal data: finding patterns of events across multiple histories. In: 2006 IEEE Symposium
    on Visual Analytics Science and Technology, pp. 167–174. IEEE (2006)
33. Forbes, A.G., Savage, S., Höllerer, T.: Visualizing and verifying directed social queries. In:
    IEEE Workshop on Interactive Visual Text Analytics, Seattle, WA (2012)
34. Gad, S., Javed, W., Ghani, S., Elmqvist, N., Ewing, T., Hampton, K.N., Ramakrishnan, N.:
    Themedelta: dynamic segmentations over temporal topic models. IEEE Trans. Vis. Comput.
    Graph. **21**(5), 672–685 (2015)
35. Gambette, P., Véronis, J.: Visualising a text with a tree cloud. In: Classification as a Tool for
    Research, pp. 561–569. Springer, Berlin (2010)
36. Gamon, M., Aue, A., Corston-Oliver, S., Ringger, E.: Pulse: mining customer opinions from
    free text. In: Advances in Intelligent Data Analysis VI, pp. 121–132. Springer, Berlin (2005)
37. Gomez-Nieto, E., San Roman, F., Pagliosa, P., Casaca, W., Helou, E.S., de Oliveira, M.C.F.,
    Nonato, L.G.: Similarity preserving snippet-based visualization of web search results. IEEE
    Trans. Vis. Comput. Graph. **20**(3), 457–470 (2014)

38. Gorg, C., Liu, Z., Kihm, J., Choo, J., Park, H., Stasko, J.: Combining computational analyses and interactive visualization for document exploration and sensemaking in jigsaw. IEEE Trans. Vis. Comput. Graph. **19**(10), 1646–1663 (2013)

39. Gschwandtner, T., Aigner, W., Kaiser, K., Miksch, S., Seyfang, A.: Carecruiser: exploring and visualizing plans, events, and effects interactively. In: Visualization Symposium (PacificVis), 2011 IEEE Pacific, pp. 43–50. IEEE (2011)

40. Guzman, E.: Visualizing emotions in software development projects. In: IEEE Working Conference on Software Visualization, pp. 1–4. IEEE (2013)

41. Hao, M.C., Rohrdantz, C., Janetzko, H., Keim, D.A., et al.: Visual sentiment analysis of customer feedback streams using geo-temporal term associations. Inf. Vis. **12**(3–4), 273 (2013)

42. Havre, S., Hetzler, B., Nowell, L.: Themeriver: visualizing theme changes over time. In: IEEE Symposium on Information Visualization, 2000. InfoVis 2000, pp. 115–123. IEEE (2000)

43. Hearst, M.A., Karadi, C.: Cat-a-cone: an interactive interface for specifying searches and viewing retrieval results using a large category hierarchy. ACM SIGIR Forum **31**(SI), 246–255 (1997)

44. Hetzler, E., Turner, A.: Analysis experiences using information visualization. IEEE Comput. Graph. Appl. **24**(5), 22–26 (2004)

45. Hinton, G.E., Roweis, S.T.: Stochastic neighbor embedding. In: Advances in neural information processing systems, pp. 833–840 (2002)

46. Hofmann, T.: Probabilistic latent semantic indexing. In: Proceedings of International ACM SIGIR Conference on Research and Development in Information Retrieval, pp. 50–57. ACM (1999)

47. Iwata, T., Yamada, T., Ueda, N.: Probabilistic latent semantic visualization: topic model for visualizing documents. In: Proceedings of SIGKDD International Conference on Knowledge Discovery and Data Mining, pp. 363–371. ACM (2008)

48. Jain, R.: Out-of-the-box data engineering events in heterogeneous data environments. In: Proceedings. 19th International Conference on Data Engineering, 2003, pp. 8–21. IEEE (2003)

49. Jänicke, S., Geßner, A., Büchler, M., Scheuermann, G.: Visualizations for text re-use. GRAPP/IVAPP, pp. 59–70 (2014)

50. Jankowska, M., Keselj, V., Milios, E.: Relative n-gram signatures: document visualization at the level of character n-grams. In: 2012 IEEE Conference on Visual Analytics Science and Technology (VAST), pp. 103–112. IEEE (2012)

51. Jolliffe, I.: Principal Component Analysis. Wiley Online Library (2002)

52. Jøsang, A.: The consensus operator for combining beliefs. Artif. Intell. **141**(1), 157–170 (2002)

53. Kaser, O., Lemire, D.: Tag-cloud drawing: algorithms for cloud visualization. arXiv preprint cs/0703109 (2007)

54. Kaster, A., Siersdorfer, S., Weikum, G.: Combining text and linguistic document representations for authorship attribution. In: SIGIR Workshop: Stylistic Analysis of Text for Information Access (2005)

55. Keim, D., Oelke, D., et al.: Literature fingerprinting: a new method for visual literary analysis. In: IEEE Symposium on Visual Analytics Science and Technology, 2007. VAST 2007, pp. 115–122. IEEE (2007)

56. Keim, D., et al.: Information visualization and visual data mining. IEEE Trans. Vis. Comput. Graph. **8**(1), 1–8 (2002)

57. Kempter, R., Sintsova, V., Musat, C., Pu, P.: Emotionwatch: visualizing fine-grained emotions in event-related tweets. In: International AAAI Conference on Weblogs and Social Media (2014)

58. Koren, Y., Carmel, L.: Visualization of labeled data using linear transformations. In: Proceedings of IEEE Symposium on Information Visualization, pp. 121–128 (2003)

59. Krstajić, M., Bertini, E., Keim, D.A.: Cloudlines: compact display of event episodes in multiple time-series. IEEE Trans. Vis. Comput. Graph. **17**(12), 2432–2439 (2011)

60. Kruskal, J.B.: Multidimensional scaling by optimizing goodness of fit to a nonmetric hypothesis. Psychometrika **29**(1), 1–27 (1964)

61. Kucher, K., Kerren, A.: Text visualization browser: a visual survey of text visualization techniques. Poster Abstracts of IEEE VIS (2014)
62. Kurby, C.A., Zacks, J.M.: Segmentation in the perception and memory of events. Trends Cogn Sci **12**(2), 72–79 (2008)
63. Le, T., Lauw, H.W.: Semantic visualization for spherical representation. In: Proceedings of the ACM SIGKDD International Conference on Knowledge Discovery and Data Mining, pp. 1007–1016. ACM (2014)
64. Lee, D.D., Seung, H.S.: Algorithms for non-negative matrix factorization. In: Advances in Neural Information Processing Systems, pp. 556–562 (2001)
65. Lee, H., Kihm, J., Choo, J., Stasko, J., Park, H.: ivisclustering: an interactive visual document clustering via topic modeling. Comput. Graph. Forum **31**(3pt3), 1155–1164 (2012)
66. Levine, N., et al.: Crimestat iii: a spatial statistics program for the analysis of crime incident locations (version 3.0). Ned Levine & Associates, Houston/National Institute of Justice, Washington (2004)
67. Lin, Y.R., Sun, J., Cao, N., Liu, S.: Contextour: contextual contour visual analysis on dynamic multi-relational clustering. In: SIAM Data Mining Conference. SIAM (2010)
68. Liu, S., Chen, Y., Wei, H., Yang, J., Zhou, K., Drucker, S.M.: Exploring topical lead-lag across corpora. IEEE Trans. Knowl. Data Eng. **27**(1), 115–129 (2015)
69. Liu, S., Wang, X., Chen, J., Zhu, J., Guo, B.: Topicpanorama: a full picture of relevant topics. In: 2014 IEEE Conference on Visual Analytics Science and Technology (VAST), pp. 183–192. IEEE (2014)
70. Liu, S., Wu, Y., Wei, E., Liu, M., Liu, Y.: Storyflow: tracking the evolution of stories. IEEE Trans. Vis. Comput. Graph. **19**(12), 2436–2445 (2013)
71. Liu, S., Zhou, M.X., Pan, S., Song, Y., Qian, W., Cai, W., Lian, X.: Tiara: interactive, topic-based visual text summarization and analysis. ACM Trans. Intell. Syst. Technol. (TIST) **3**(2), 25 (2012)
72. Luo, D., Yang, J., Krstajic, M., Ribarsky, W., Keim, D.: Eventriver: visually exploring text collections with temporal references. IEEE Trans. Vis. Comput. Graph. **18**(1), 93–105 (2012)
73. Van der Maaten, L., Hinton, G.: Visualizing data using t-sne. J. Mach. Learn. Res. **9**(2579–2605), 85 (2008)
74. Makki, R., Brooks, S., Milios, E.E.: Context-specific sentiment lexicon expansion via minimal user interaction. In: Proceedings of the International Conference on Information Visualization Theory and Applications (IVAPP), pp. 178–186 (2014)
75. Marcus, A., Bernstein, M.S., Badar, O., Karger, D.R., Madden, S., Miller, R.C.: Twitinfo: aggregating and visualizing microblogs for event exploration. In: Proceedings of the SIGCHI Conference on Human Factors in Computing Systems, pp. 227–236. ACM (2011)
76. Miller, N.E., Wong, P.C., Brewster, M., Foote, H.: Topic islands TM—a wavelet-based text visualization system. In: IEEE Symposium on Information Visualization, pp. 189–196. IEEE (1998)
77. Munroe, R.: Movie narrative charts. http://xkcd.com/657/. Accessed Jan 2016
78. Neto, J.L., Santos, A.D., Kaestner, C.A., Freitas, A.A.: Document clustering and text summarization. In: Proceedings of the International Conference Practical Applications of Knowledge Discovery and Data Mining, pp. 41–55. The Practical Application Company (2000)
79. Oelke, D., Hao, M., Rohrdantz, C., Keim, D., Dayal, U., Haug, L.E., Janetzko, H., et al.: Visual opinion analysis of customer feedback data. In: IEEE Symposium on Visual Analytics Science and Technology, 2009. VAST 2009, pp. 187–194. IEEE (2009)
80. Oelke, D., Kokkinakis, D., Keim, D.A.: Fingerprint matrices: uncovering the dynamics of social networks in prose literature. Comput. Graph. Forum **32**(3pt4), 371–380 (2013)
81. Oelke, D., Strobelt, H., Rohrdantz, C., Gurevych, I., Deussen, O.: Comparative exploration of document collections: a visual analytics approach. Comput. Graph. Forum **33**(3), 201–210 (2014)
82. Ogawa, M., Ma, K.L.: Software evolution storylines. In: Proceedings of the 5th International Symposium on Software Visualization, pp. 35–42. ACM (2010)
83. Ogievetsky, V.: PlotWeaver. http://ogievetsky.com/PlotWeaver/. Accessed Jan 2016

84. Pascual-Cid, V., Kaltenbrunner, A.: Exploring asynchronous online discussions through hierarchical visualisation. In: 2009 13th International Conference on Information Visualisation, pp. 191–196. IEEE (2009)

85. Paulovich, F.V., Minghim, R.: Hipp: a novel hierarchical point placement strategy and its application to the exploration of document collections. IEEE Trans. Vis. Comput. Graph. **14**(6), 1229–1236 (2008)

86. Plaisant, C., Mushlin, R., Snyder, A., Li, J., Heller, D., Shneiderman, B.: Lifelines: using visualization to enhance navigation and analysis of patient records. In: Proceedings of the AMIA Symposium, p. 76. American Medical Informatics Association (1998)

87. Reisinger, J., Waters, A., Silverthorn, B., Mooney, R.J.: Spherical topic models. In: Proceedings of International Conference on Machine Learning, pp. 903–910 (2010)

88. Riehmann, P., Gruendl, H., Potthast, M., Trenkmann, M., Stein, B., Froehlich, B.: Wordgraph: keyword-in-context visualization for netspeak's wildcard search. IEEE Trans. Vis. Comput. Graph. **18**(9), 1411–1423 (2012)

89. Robertson, G., Czerwinski, M., Larson, K., Robbins, D.C., Thiel, D., Van Dantzich, M.: Data mountain: using spatial memory for document management. In: Proceedings of the 11th Annual ACM Symposium on User Interface Software and Technology, pp. 153–162. ACM (1998)

90. Robertson, G.G., Mackinlay, J.D.: The document lens. In: Proceedings of the 6th Annual ACM Symposium on User Interface Software and Technology, pp. 101–108. ACM (1993)

91. Rohrdantz, C., Hao, M.C., Dayal, U., Haug, L.E., Keim, D.A.: Feature-based visual sentiment analysis of text document streams. ACM Trans. Intell. Syst. Technol. (TIST) **3**(2), 26 (2012)

92. Roweis, S.T., Saul, L.K.: Nonlinear dimensionality reduction by locally linear embedding. Science **290**(5500), 2323–2326 (2000)

93. Rusu, D., Fortuna, B., Mladenić, D., Grobelnik, M., Sipos, R.: Document visualization based on semantic graphs. In: 13th International Conference Information Visualisation, pp. 292–297. IEEE (2009)

94. Sebrechts, M.M., Cugini, J.V., Laskowski, S.J., Vasilakis, J., Miller, M.S.: Visualization of search results: a comparative evaluation of text, 2d, and 3d interfaces. In: Proceedings of the 22nd Annual International ACM SIGIR Conference on Research and Development in Information Retrieval, pp. 3–10. ACM (1999)

95. Smith, D.A.: Detecting events with date and place information in unstructured text. In: Proceedings of the 2nd ACM/IEEE-CS Joint Conference on Digital Libraries, pp. 191–196. ACM (2002)

96. Spoerri, A.: Infocrystal: a visual tool for information retrieval & management. In: Proceedings of the Second International Conference on Information and Knowledge Management, pp. 11–20. ACM (1993)

97. Stasko, J., Zhang, E.: Focus+ context display and navigation techniques for enhancing radial, space-filling hierarchy visualizations. In: IEEE Symposium on Information Visualization, 2000. InfoVis 2000, pp. 57–65. IEEE (2000)

98. Stoffel, A., Strobelt, H., Deussen, O., Keim, D.A.: Document thumbnails with variable text scaling. Comput. Graph. Forum **31**(3pt3), 1165–1173 (2012)

99. Strobelt, H., Oelke, D., Rohrdantz, C., Stoffel, A., Keim, D., Deussen, O., et al.: Document cards: a top trumps visualization for documents. IEEE Trans. Vis. Comput. Graph. **15**(6), 1145–1152 (2009)

100. Tanahashi, Y., Hsueh, C.H., Ma, K.L.: An efficient framework for generating storyline visualizations from streaming data. IEEE Trans. Vis. Comput. Graph. **21**(6), 730–742 (2015)

101. Tanahashi, Y., Ma, K.L.: Design considerations for optimizing storyline visualizations. IEEE Trans. Vis. Comput. Graph. **18**(12), 2679–2688 (2012)

102. Thai, V., Handschuh, S., Decker, S.: Tight coupling of personal interests with multidimensional visualization for exploration and analysis of text collections. In: International Conference on Information Visualisation, pp. 221–226. IEEE (2008)

103. Van Ham, F., Wattenberg, M., Viégas, F.B.: Mapping text with phrase nets. IEEE Trans. Vis. Comput. Graph. **15**(6), 1169–1176 (2009)

104. Viegas, F.B., Wattenberg, M., Feinberg, J.: Participatory visualization with wordle. IEEE Trans. Vis. Comput. Graph. **15**(6), 1137–1144 (2009)
105. Wang, C., Xiao, Z., Liu, Y., Xu, Y., Zhou, A., Zhang, K.: Sentiview: sentiment analysis and visualization for internet popular topics. IEEE Trans. Hum. Mach. Syst **43**(6), 620–630 (2013)
106. Wang, T.D., Plaisant, C., Shneiderman, B., Spring, N., Roseman, D., Marchand, G., Mukherjee, V., Smith, M.: Temporal summaries: supporting temporal categorical searching, aggregation and comparison. IEEE Trans. Vis. Comput. Graph. **15**(6), 1049–1056 (2009)
107. Wanner, F., Rohrdantz, C., Mansmann, F., Oelke, D., Keim, D.A.: Visual sentiment analysis of rss news feeds featuring the us presidential election in 2008. In: Workshop on Visual Interfaces to the Social and the Semantic Web (VISSW) (2009)
108. Wattenberg, M.: Arc diagrams: visualizing structure in strings. In: IEEE Symposium on Information Visualization, pp. 110–116. IEEE (2002)
109. Wattenberg, M., Viégas, F.B.: The word tree, an interactive visual concordance. IEEE Trans. Vis. Comput. Graph. **14**(6), 1221–1228 (2008)
110. Wensel, A.M., Sood, S.O.: Vibes: visualizing changing emotional states in personal stories. In: Proceedings of the 2nd ACM International Workshop on Story Representation, Mechanism and Context, pp. 49–56. ACM (2008)
111. Wongsuphasawat, K., Gotz, D.: Exploring flow, factors, and outcomes of temporal event sequences with the outflow visualization. IEEE Trans. Vis. Comput Graph. **18**(12), 2659–2668 (2012)
112. Wongsuphasawat, K., Guerra Gómez, J.A., Plaisant, C., Wang, T.D., Taieb-Maimon, M., Shneiderman, B.: Lifeflow: visualizing an overview of event sequences. In: Proceedings of the SIGCHI Conference on Human Factors in Computing Systems, pp. 1747–1756. ACM (2011)
113. Wongsuphasawat, K., Shneiderman, B.: Finding comparable temporal categorical records: a similarity measure with an interactive visualization. In: IEEE Symposium on Visual Analytics Science and Technology, 2009. VAST 2009, pp. 27–34. IEEE (2009)
114. Wu, Y., Provan, T., Wei, F., Liu, S., Ma, K.L.: Semantic-preserving word clouds by seam carving. Comput. Graph. Forum **30**(3), 741–750 (2011)
115. Wu, Y., Wei, F., Liu, S., Au, N., Cui, W., Zhou, H., Qu, H.: Opinionseer: interactive visualization of hotel customer feedback. IEEE Trans. Vis. Comput. Graph. **16**(6), 1109–1118 (2010)
116. Zacks, J.M., Tversky, B.: Event structure in perception and conception. Psychol. Bull. **127**(1), 3 (2001)
117. Zhang, C., Liu, Y., Wang, C.: Time-space varying visual analysis of micro-blog sentiment. In: Proceedings of the 6th International Symposium on Visual Information Communication and Interaction, pp. 64–71. ACM (2013)
118. Zhao, J., Gou, L., Wang, F., Zhou, M.: Pearl: an interactive visual analytic tool for understanding personal emotion style derived from social media. In: 2014 IEEE Conference on Visual Analytics Science and Technology (VAST), pp. 203–212. IEEE (2014)

Chapter 3
Data Model

Abstract In text data, the semantic relationships among keywords, sentences, paragraphs, sections, chapters, and documents are usually implicit. A reader must go through the entire corpus to capture insights such as relationships among characters in a novel, event causalities in news reports, and evolution of topics in research articles. The difficulties usually arise from the unstructured nature of text data as well as the low information acquisition efficiency of reading these unstructured texts. Therefore, how to convert unstructured text data into a structured form to facilitate understanding and cognition becomes an important problem that has attracted considerable research interest. In this chapter, we introduce the data models that are frequently used in current text visualization techniques. We review low-level data structures such as bag of words, the structures at the syntactic level such as the syntax tree, as well as the network-oriented data structures at the semantic level. We introduce these data models (i.e., structures) together with detailed visualization examples that show how the structures are used to represent and summarize the unstructured text data.

Converting unstructured documents into a structured form helps produce additional context regarding data relationships, thereby further helping users to quickly capture the insight of the document content as well as complex relationships among keywords, paragraphs, and sections in the documents at the semantic level. For example, in the field of data visualization, Watternberg et al. [14] introduced WordTree (Fig. 2.8), which used a tree structure to decompose sentences into text to investigate how a collection of sentences extracted from the documents share similar structures. Van Ham et al. [12] introduced PhraseNets (Fig. 2.9), which employs a graph based representation to capture co-occurrences of keywords based on textual patterns such as "X of Y" or "A's B" in document collections. Based on these techniques, many interesting patterns are illustrated, helping users to summarize, reason, compare, and memorize the content and the corresponding semantics of the input text data. In addition, organizing and storing documents in structured form also aids to the data analysis process. Many advanced analysis algorithms such as clustering and classification are developed to deal with structured data described by feature vectors and relationships. Given all these factors, converting a collection of text data into a structured data model is one of the most important steps for text analysis and visualization.

© Atlantis Press and the author(s) 2016

C. Nan and W. Cui, *Introduction to Text Visualization*, Atlantis Briefs
in Artificial Intelligence 1, DOI 10.2991/978-94-6239-186-4_3

However, designing a structured data model to capture the insight structures of text data is not an easy task. Many challenges exist. First, the text data itself contains rich information that can be difficult in designing a comprehensive model to capture all the information at the same time. For example, in a digital publication dataset, each publication is a document that contains information on authors, content in different topics, and publishing venue and time. Thus, different publications in a dataset may be related with each other from different aspects in terms of topics, authors, and the corresponding venues and time. Capturing all these information at the same time is not an easy task. Second, the text data may dynamically changed overtime, thus capturing the corresponding information dynamics in the model is challenging. Third, the data corpus could be extremely large, thereby requiring a scalable model to store the data. For example, Twitter produces millions or even billions of tweet everyday which are large and dynamic.

To address these challenges, a well-designed data model should consider many issues to capture all kinds of key attributes of the text data at the same time. First, as a fundamental requirement, the model should be able to store the content of the inputting documents. Second, to facilitate data understanding and reasoning, the model should capture the innate relationships of the key elements extracted from the text. These elements could be the representative keywords captured in the topic models [1], or the name entities detected in the process of name entity recognition [10]. These textual elements are usually related to each other in different ways such as through co-occurrences in documents or paragraphs that belonging to the same topic, or through semantic similarities. Third, in order to further clarify the content for summarization purpose, the model should also be able to capture both content and relationships from different information facets to produce a mutli-perspective representation of the data. With these considerations in mind, next, we will introduce data models that are widely used for text analysis and visualizations.

Based on the aforementioned considerations, many data models and structures have been developed for text data analysis and visualization, including the simplest Bag-of-Words model,[1] more advanced tree and graph models, and the sophisticated multifaceted entity-relational data model [3, 4]. In the following subsections, we will describe the details of these data models one by one to introduce the related concepts, their advantages and disadvantages, as well as the corresponding visualization examples based on them.

---

[1]Bag of words model. http://en.wikipedia.org/wiki/Bag-of-words_model.

3.1 Data Structures at the Word Level

3.1.1 Bag of Words and N-Gram

This is the simplest representation of documents, which is widely used in natural language processing and information retrieval (IR). In this model, a text (such as a sentence or a document) is represented as the bag (multiset) of its words, disregarding grammar and even word order but maintaining multiplicity. The bag-of-words model is commonly used in methods of document classification methods where the (frequency of) occurrence of each word is used as a feature to train a classifier. Documents modeled by bag of words can be directly visualized by tag clouds or Wordle [13] as illustrated in Fig. 2.7.

In recent years, growing attention has focused on analyzing and visualizing streaming text data produced by various media systems such as online news, microblogs, and email exchanges. To dealing with these data, the most commonly used approach is to produce a dynamic bag-of-words model in which the words are changed over time corresponding to the change of the underlying text. The most famous visualization based on this data model is the ThemeRiever and many of its variants [7, 8] as shown in Fig. 2.12.

The Bag-of-Words data model is the simplest and the most intuitive representation of the text data. When the proper words are extracted, this model successfully summarizes the content of the text. However, it usually fails to capture the semantic relationships among different words. Intuitively, in the tag cloud visualization, the words are usually randomly laid out or following some heuristic strategies such as putting the high frequency words in the middle surrounded by low frequency ones. In these visualizations, the screen distances between words usually are meaningless, i.e., two nearby wards do not mean they are not necessarily similar or related to each other, thereby making the visualization sometimes difficult to read and understand. To solve this problem, more advanced tree and graph models are developed, which are introduced in the following.

3.1.2 Word Frequency Vector

When computing the feature vectors of each document, usually, only the terms that best differentiate various documents are selected. The term extraction usually follows a standard procedure in which the documents are first split into words, and then a set of predefined stop words are removed and finally the remaining words are ranked based on their importance and computed based on their frequencies. In particular, TF-IDF is the most commonly used method to rank the words, which ensures that the selected high-ranking words are able to differentiate the documents. In this case, TF is the term frequency, which measures how frequently a term t occurs in a document

d, which is usually normalized by document length (i.e., the total number of terms in the document). Formally, TF is defined as:

$$TF(t, d) = freq(t, d)/lens(d)$$

where $freq(\cdot)$ computes the term frequency in the given document d and $len(\cdot)$ indicates the document length. IDF refers to the inverse document frequency, which measures how important a term is in the document collection, which is formally defined as:

$$IDF(t, D) = ln(|D|/|\{t \in d | d \in D\}|)$$

where D is the document collection. Intuitively, IDF weights down the frequent terms that widely occur in many documents such as "this" and "that", but scale up the ones that high frequently exist in one of some of the documents, thereby ranking out the words that best differentiate the documents. The TF-IDF score is defined as

$$TF\text{-}IDF(t, d, D) = TF(t, d) \cdot IDF(t, D)$$

Sometimes, stemming is applied to extract word radicals so that the frequency can be more correctly computed and an n-gram is computed to increase the amount of information of each term before the ranking procedure.

3.2   Data Structures at the Syntactical-Level

In text analysis, the tree structure has been widely used to capture the relationships of the text content. It is generally used in two approaches. First, it is used in the procedure of hierarchical text classification [11], and topic modeling [6] to produce document classes or topic clusters with multiple granularity levels. Analysis in this category, usually accompanied with a zoomable representation of the results, allows users to interactively zooming into different levels of information detail. In these applications, the tree data model is used to guide the interactive data navigation. Second, in the field of nature language processing, the tree data model is also used for syntactic analysis [9]. A corresponding visualization, Word Tree [14] is also designed for this purpose, which represents the syntactic structure of a collection of sentences (Fig. 2.8).

Compared with the Bag-of-Words model, the tree structure is advanced and is able to guide the navigation and capture the structure of sentences. However, the higher level relationships, i.e., the relationships at the semantic level, are still obscure in the tree model. For example, various relationships among different name entities, such as authors and topic keywords in a publication dataset, can not be clearly captured in a tree structure.

3.3 Data Models at the Semantic Level

3.3.1 Network Oriented Data Models

To address the aforementioned limitations, various types of network oriented structures are frequently used to capture the semantics of the text data, in which nodes (i.e., entities) indicate concepts, name entities, or keywords, showing the content of corpus and links (i.e., relation) indicates the their relationships. It is one of the most flexible structures designed to capture relationships among data elements. In text analysis, the network oriented data model is widely used to represent high level relationships among documents such as citing and referencing among publications and hyperlinks in online webpages. This data model can also be used to capture the textual relationships at the content level by representing the relationships among topic keywords or name entities that are extracted from the documents. Phrase Nets is a visualization which is designed based on the network data model to illustrate relationships among keywords. In this visualization, a user can choose relationships based on different grammar pattern. For example, "A $\acute{s}$ B" or "B of A" indicates the relationship of B belonging to A. Once the pattern is chosen (Fig. 2.9), the text is tokenized by "'" or "of" and the high frequency keywords in the text and their corresponding relationships are extracted which are visualized in a graph visualization.

Recently, a directed network model is introduced to capture the evolution of topics over time. As shown in Fig. 3.1, in this model, each node represents a topic computed based on the underlying text data that are collected at different time. In this case, the nodes from left to right indicate topics respectively computed at time t_1, t_2, and t_3. The directed links indicate the transition with weights indicating the corresponding portion of the content transited from one topic to another topic. Obviously, this data model focuses on capturing the transition trend, i.e., how topics are merged or split over time instead of the detailed changes of the keywords. For example, in Fig. 3.1, the topics **T1-1** and **T1-2** at t_1 are merged into a new topic **T2-1** at time t_2. When visualizing this directed network model in TextFlow [5] as illustrated in Fig. 2.13, the dynamic topic transition trend are clearly shown.

Fig. 3.1 The directed network data model for modeling topic evolutions over time

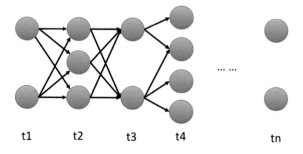

t1   t2   t3   t4   tn

3.3.2 Multifaceted Entity-Relational Data Model

Although flexible, the aforementioned network oriented data models are fail to capture both content and relationships from different information facets to produce a mutli-perspective representation of the data. To meet this requirement, the multifaceted entity-relational data model is developed based on the network data model. This model is a more sophisticated design that decomposes text data into several key elements from different aspects to facilitate analysis and visualization. This model is used in most of the algorithms and visualizations introduced in this book and is the most comprehensive as well as the most complicated data model for representing text data among all the models that we introduced.

Generally speaking, the Multifaceted Entity-Relational data model (as shown in Fig. 3.2) decomposes the text data into three low-level components including entities, relations, and facets based on which the data model reveals high level patterns such as clusters, topics, and the corresponding transition trend of the text data over time. Specifically, we interpret the key components and the corresponding high level patterns as follows:

- **Entities** are instances (e.g. keywords or name entities) of a particular concept (e.g., topics) extracted from the underlying text data. An entity is considered as the atom element in the model that cannot be further divided.
- **Facets** are classes of entities that share a unique attribute or attributes. For example, "topic" is a facet that contains a set of topic keywords (i.e., entities) together describing a unique semantic meaning.
- **Relations** are connections between pairs of entities. In the Multifaceted-Entity Relational Data model, there are two types of relations. *Internal relations* are connections between entities or entity groups within the same information facet. An example is the co-occurrence relationships of the keywords within the same topic. *External relations* are connections between entities of different facets. For example the co-occurrence relationships of the keywords cross different topics.

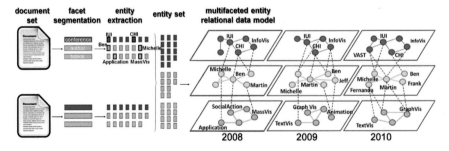

Fig. 3.2 The multifaceted entity relational data model and the data transformation pipeline. The DBLP data are used as an example

- **Clusters** are groups of similar entities within a single facet. It is a concept that further divides entities within the same facet into groups based on their similarities such as sub-topics under a primary topic.
- **Temporal Trend** describes the information dynamics of both entities and relationships over time. As shown in Fig. 3.2, this data model captures the change of the data by producing the multifaceted entity-relational structures based on the data collected at different times.

An example of using the data model to store a publication dataset is illustrated in Fig. 3.2. In this example, the publications are divided into three information facets—Conferences (where the paper published in), Authors, and Keywords (describe the topics of the papers)—each represented as a separate layer in Fig. 3.2. Nodes on each layer represent entities within the corresponding facet. Edges within a layer are internal relations, while edges across layers are external relations. This model also captures the annual changes on the dataset.

Data Transformation. Compared with the aforementioned data models, the Multifaceted Entity-Relational data model is relatively complex. Transforming text data into this model usually requires to performing an entire data preprocessing procedure that consists of multiple key steps, including facet segmentation, entity extraction, relation building, and temporal reordering.

- **Facet Segmentation**. In this step, documents are divided into information facets. Usually, no standard approach for doing this exists because it highly depends on the input documents. Sometimes, the information facets can be identified easily. In the aforementioned example of the publication dataset, conference, author, and keywords are three obvious information facets that can be easily extracted from the data. Segmenting online descriptions of diseases that are usually written in a standard format, as well as describing the corresponding symptoms, diagnosis, and treatment are also easy. However, in most cases, the documents are not well organized in a standard format. To segment these data, highly advanced techniques need to be used. Typically, we employ a topic modeling technique such as LDA [2] to compute topics of the input text and to take each topic as a facet.
- **Entity Extraction**. Entities are further extracted from each of the information facet identified in the above procedure. These entities can be keywords ranked by topic models or name-entities identified in a name entity recognition procedure.
- **Relation Building**. In this step, we establish relationships among entities within or across different information facets. Among different entities, there are many types of relationships such as co-occurrences in documents, or various relationships (e.g., A ś B or B of A, where A and B are textual entities) in sentences.
- **Temporal Ordering**. In the final step of the data transformation, all the data entities and their corresponding relations are sorted in the temporal order, thus producing a data stream that can be further segmented by time windows with either fixed or adaptive sizes.

3.4   Summary of the Chapter

In this chapter, we introduced the data models and the corresponding methods to convert the unstructured text data into a structured form to facilitate text analysis and visualization. We first introduce the background of the problem, including motivations and corresponding considerations and requirements. Thereafter, we discuss four most frequently used data models, namely, bag-of-words, tree network, and multifaceted entity relational data model to represent text data in a structured form. We describe these models one by one in detail and introduce the related concepts, their advantages and disadvantages, as well as the corresponding visualization examples. Among all these data models, the last, which is the most comprehensive one that enables a multifaceted representation of the text data, is the focus of our study. This model is also the foundation of most of the algorithms and visualizations introduced in this book. Therefore, gaining an in-depth understanding of this data model is important in reading the following chapters.

References

1. Blei, D.M.: Probabilistic topic models. Commun. ACM **55**(4), 77–84 (2012)
2. Blei, D.M., Ng, A.Y., Jordan, M.I.: Latent Dirichlet allocation. J. Mach. Learn. Res. **3**, 993–1022 (2003)
3. Cao, N., Gotz, D., Sun, J., Lin, Y.R., Qu, H.: Solarmap: multifaceted visual analytics for topic exploration. In: IEEE International Conference on Data Mining, pp. 101–110. IEEE (2011)
4. Cao, N., Sun, J., Lin, Y.R., Gotz, D., Liu, S., Qu, H.: Facetatlas: multifaceted visualization for rich text corpora. IEEE Trans. Vis. Comput. Graph. **16**(6), 1172–1181 (2010)
5. Cui, W., Liu, S., Tan, L., Shi, C., Song, Y., Gao, Z.J., Qu, H., Tong, X.: Textflow: towards better understanding of evolving topics in text. IEEE Trans. Vis. Comput. Graph. **17**(12), 2412–2421 (2011)
6. Griffiths, D.: Tenenbaum: hierarchical topic models and the nested Chinese restaurant process. Adv. Neural Inf. Process. Syst. **16**, 17 (2004)
7. Havre, S., Hetzler, B., Nowell, L.: Themeriver: visualizing theme changes over time. In: IEEE Symposium on Information Visualization, 2000. InfoVis 2000, pp. 115–123. IEEE (2000)
8. Liu, S., Zhou, M.X., Pan, S., Song, Y., Qian, W., Cai, W., Lian, X.: TIARA: interactive, topic-based visual text summarization and analysis. ACM Trans. Intell. Syst. Technol. (TIST) **3**(2), 25 (2012)
9. Manning, C.D., Schütze, H.: Foundations of Statistical Natural Language Processing, vol. 999. MIT Press, Cambridge (1999)
10. Nadeau, D., Sekine, S.: A survey of named entity recognition and classification. Lingvist. Investig. **30**(1), 3–26 (2007)
11. Sun, A., Lim, E.P.: Hierarchical text classification and evaluation. In: IEEE International Conference on Data Mining, pp. 521–528. IEEE (2001)
12. Van Ham, F., Wattenberg, M., Viégas, F.B.: Mapping text with phrase nets. IEEE Trans. Vis. Comput. Graph. **15**(6), 1169–1176 (2009)
13. Viegas, F.B., Wattenberg, M., Feinberg, J.: Participatory visualization with wordle. IEEE Trans. Vis. Comput. Graph. **15**(6), 1137–1144 (2009)
14. Wattenberg, M., Viégas, F.B.: The word tree, an interactive visual concordance. IEEE Trans. Vis. Comput. Graph. **14**(6), 1221–1228 (2008)

Chapter 4
Visualizing Document Similarity

Abstract A large category of text visualization techniques were developed to illustrate similarities of document files in a corpus. This is the most traditional research direction in this field. These techniques produce visualizations in a similar form in which document files are represented as points on the display sized by their importance and colored or shaped by their semantics (such as topics). The screen distance between any pair of points indicates the similarities of the corresponding documents that are captured by different measures in different analysis models, following the rule of the closer, the more similar. Two major types of approaches: projection (i.e., dimension reduction)-based methods and semantic-oriented document visualizations, are proposed to produce such a document overview; these approaches are developed based on different data and analysis models. In this section, we investigate these techniques and the corresponding document visualization systems.

4.1 Projection Based Approaches

One of the major method for visualizing document similarity is the projection based approaches. In these techniques, a document is represented by an N-dimensional feature vector in which each filed represents the number of a term in a keyword collection extracted from the entire document corpus. The visualization is produced by projecting the document from the high dimensional feature space into a low dimensional (2D or 3D) visualization space.

Formally, a projection procedure is defined as a follows: let $X = \{x_1, x_2, \ldots, x_k\}$ be a set of n-dimensional data items (i.e., documents), with $d_n(x_i, x_j)$ as a defined distance measure between two items x_i and x_j. Let $Y = \{y_1, y_2, \ldots, y_n\}$ also be a set of points defined in an m-dimensional space, with $M \in \{1, 2, 3\}$, and $d_m(x_i, x_j)$ indicates the distance between two data items i and j in the m-dimensional visualization space. In this way, a dimension reduction technique is described as an injective function $f : X \rightarrow Y$ that attempts to best preserve a specified relationship between d_n and d_m. This definition treats each projection as an optimization problem that has two advantages [19]: (1) the iterative nature of the optimization allows users to observe the details of the projection process, and (2) the incremental nature of the

C. Nan and W. Cui, *Introduction to Text Visualization*, Atlantis Briefs in Artificial Intelligence 1, DOI 10.2991/978-94-6239-186-4_4

optimization allows users to add new data items efficiently. Many such projection techniques have been described in the literature. Overall, they can be classified into two major categories, based on whether function f is linear or non-linear.

In the following, we first describe the methods developed for selecting proper features for each document, followed by a brief introductions of both linear and non-linear projection techniques.

4.1.1 Linear Projections

These techniques project high dimensional data based on linear combinations of the dimensions, sharing many advantages than non-linear projections [14]. These advantages include the following: (1) it is assured for linear projections to show genuine properties of the data, thereby making them reliable; (2) the linear combinations of the original axes make the new axes generated by projection meaningful; (3) the technique is an incremental process in which new data items can be easily added based on the already computed linear transformation; (4) computing a linear projection is easy and efficient. Thus, the linear projection attracts serious attentions in the field of information visualization and many of them have been used to represent the overview of document similarities.

Koren and Kermal [14] first introduced the general form of linear projection into the visualization community. In particular, the linear projection can be formulated as the following optimization problem:

$$max \sum_{i<j} d_m(x_i, x_j)^2$$

which seeks an m-dimensional projection that best maximizes the distances between every two points x_i, x_j in the n-dimensional space. Choosing different distance metrics will result in different algorithms. In this study, we show two of them, Principle Component Analysis and Linear Discriminant Analysis, as examples. More discussions can be found in [14].

Principal Component Analysis (PCA). When setting $d_m = ||x_i - x_j||$ (i.e., the L2-norm) in the aforementioned model, we have

$$max \sum_{i<j} ||x_i - x_j||^2$$

which is the objective of *Principal Component Analysis (PCA)* [11], a well-know linear dimension reduction technique. Intuitively, this technique aims to preserve and maximize data variance so that each point in the data can be easily differentiated from others when shown on the screen while the projection is being computed. This optimization problem can be effectively solved by decomposing the covariance matrix $\frac{1}{n}X^T X$ of the document collection ($X = \{x_1, \ldots, x_k\}$ into orthogonal eigenvectors and taking the first m eigenvectors $y_1, y_2, \ldots, y_p$ with the largest eigenvalues as the m-dimensional visualization coordinates. Following this definition, the so-called weighted PCA is defined as:

$$max \sum_{i<j} \omega_{ij} ||x_i - x_j||^2$$

where ω_{ij} measures the importance of separating two data items x_i and x_j. It has many variances designed for different purpose.

For example, when $\omega_{ij} = 1/d_n(x_i, x_j)$, the preceding objective computes a normalized PCA [11] that transforms the distances computed under different features into the same scale, thereby improving the robustness of the algorithm.

When setting $\omega_{ij} = t \cdot \omega_{ij}$, the aforementioned model is able to represent and interpret labeled multidimensional data (e.g., classification results or documents labeled with different topic tags in the context of text visualization) by adjusting the value of t. In particular, setting $0 \le t \le 1$ if both x_i and x_j share the same label (i.e., in the same class) otherwise setting $t = 1$ if x_i, x_j are in different classes, helps to differentiate different data classes in the resulting visualization.

Linear Discriminant Analysis (LDA). When applying the following distance metric to the above linear projection model, the resulting objective indicates of Fisher's *Linear Discriminant Analysis (LDA)* [2], another type of linear projection that is developed to illustrate the labeled multivariate data (e.g., classification results).

$$d_m(i, j) = \begin{cases} \frac{1}{k^2} - \frac{1}{k \cdot |C|} & \text{if item i and j are both in the class C} \\ \frac{1}{k^2} & \text{if item i and j are in different classes} \end{cases}$$

where k indicates the total number of data items in the collection and C indicates a class whose size (i.e., the number of containing items) is $|C|$. Similar to the above weighted PCA with the weight determining the label information, this distance metric also helps to separate different classes in the data and considers the size of each class.

4.1.2 Non-linear Projections

The aforementioned linear projection techniques, although efficient and has many advantages, fail to capture the higher order relationship of the data items that are usually non-linear. To address this issue, many non-linear projection techniques are also introduced [9, 15, 17, 20, 21]. Based on the ways in which data relationships are measured, these techniques can be largely organized into two categories including, (1) those that use pairwise distances such as multidimensional scaling (MDS) [15], locally linear embedding (LLE) [20] and isomap [21] and (2) those that employ probabilistic formulations such as stochastic neighbor embedding (SNE) [7], parametric embedding (PE) [9] and t-SNE [17]. Next, we will introduce the one representative technique in each category as follows.

Distance Oriented Techniques

Multidimensional Scaling (MDS) [15] is the most well-known distance oriented projection technique that has been commonly used for visualizing multidimensional data. It focuses on placing data items (i.e., documents in case of text visualization) in the m-dimensional (m = 2 or 3) visualization space so that the high-dimensional distance between any pair of data items is well preserved. MDS achieves this goal by optimizing the following objective:

$$min \sum_{i<j} ||d_m(x_i, x_j) - d_n(x_i, x_j)||^2$$

where $d_m(x_i, x_j)$ and $d_n(x_i, x_j)$, which is consistent with the previous definition, respectively indicates the screen distance and feature distance between the data items i and j. Usually, L2-norm is used. In other sophisticated usages of MDS such as Isomap [3], the connections (i.e., relationships) between data points are estimated based on the nearest Euclidean neighbors in high-dimensional space and the distance is thus defined based on the shortest paths in the resulting graph.

MDS usually generates meaningful results, revealing patterns such as clusters and outliers. Therefore, it has been widely used in the field of data visualization to produce an overview of the entire document corpus [1]. However, achieving the aforementioned objective requires calculating the pairwise distance of data points, which is time consuming with a time complexity of $O(n^2)$. To address this issue, many approximation methods have been developed [5, 12, 18, 22], which introduces methods for computing approximate MDS projection results with a linear or near-linear time complexity. However, in these techniques, the projection quality is sacrificed.

The preceding objective can also be generalized in the following weighted form:

$$min \sum_{i<j} \omega_{ij} ||d_m(x_i, x_j) - d_n(x_i, x_j)||^2$$

where ω_{ij} indicates the importance of preserving the distance between a specified pair of data items i and j in the low dimensional space. A different ω_{ij} leads to different projection models. In particular, when setting $\omega_{ij} = 1/d_m^k$ where $k \in [1, m]$, it becomes the Kamada-Kawai force-directed layout [13], a well-known graph drawing algorithm that can be bettered solved based on stress majorization [6].

Probabilistic Formulation

Among all probabilistic formulation oriented non-linear projection techniques such as stochastic neighbor embedding (SNE) [7] and parametric embedding (PE) [9], t-SNE [17], a variant of SNE technique, is considered as a landmark technique that produces the best visualization results in most cases, which preserves the details of the data's local structure in the low dimensional visualization space. Compared with SNE, t-SNE overcomes the crowding problem and produces a visualization that effectively reveals cluster patterns.

Specifically, SNE minimizes the sum of Kullback-Leibler divergences [16] over all data points via the gradient descent method as follows:

$$min \sum_i \sum_j p_{j|i} log \frac{p_{j|i}}{q_{j|i}}$$

where $p_{j|i}$ and $q_{j|i}$ are conditional probabilities that capture the similarity between data point x_i and x_j in high dimensional feature space and low dimensional visualization space respectively. Intuitively, they can be understood as when given data point x_i, the probability of choosing x_j as its neighbor in the feature space $(p_{j|i})$ and visualization space $(q_{j|i})$. In an ideal projection in which all the features are preserved, $p_{j|i}$ and $q_{j|i}$ should be the same, which provides the design rationale of SNE: it computes the projection by minimizing the differences (measured by Kullback-Leibler divergence in the aforementioned optimization objective) between these two probabilities over all data points.

Although the preceding optimization problem is well motivated and defined, the visualization results produced by SNE usually fail to capture data clusters, which is known as the crowding problem [17]. t-SEN addresses this issue from two aspects: first, it employs the joint probability p_{ij} and q_{ij} instead of the conditional probability to provide a symmetric measurement of the similarity between data point x_i and x_j in both the feature and visualization space. Second, t-SEN defines q_{ij} based on the Student's t-distributions instead of Gaussian distribution. As shown in Fig. 4.1, the resulting model:

Fig. 4.1 Visualizations of 6000 handwritten digits from the MNIST dataset based on SNE (*left*) and t-SEN (*right*)

$$min \sum_i \sum_j p_{ij} log \frac{p_{ij}}{q_{ij}}$$

better preserves local structures and successfully preserves the high-dimensional data clusters in the visualization.

4.2   Semantic Oriented Techniques

Unlike the aforementioned projection-based visualization approaches, semantic-oriented techniques represent document similarity based on latent topic information. Research in this direction is inspired by traditional topic modeling such as PLSA [8] and Latent Dirichlet allocation (LDA) [4]. Although widely used and is able to extract a low dimensional representation of a document, these techniques can not represent more than three topics in 2D visualization space as the visualization has to be produced inside a simplex space, which is usually shown as a triangle on the 2D Euclidean visualization plane as shown in Fig. 4.2d. To effectively visualize document similarity based on multiple topics, many semantic-oriented techniques are developed. Probabilistic Latent Semantic Visualization (PLSV) [10] is the first and the most impactful work in this direction. Compared with PLSA and LDA, PLSV can represent any number of topics within a two-dimensional display, and it embeds documents in the Euclidean space instead of the simplex space, thus produc-

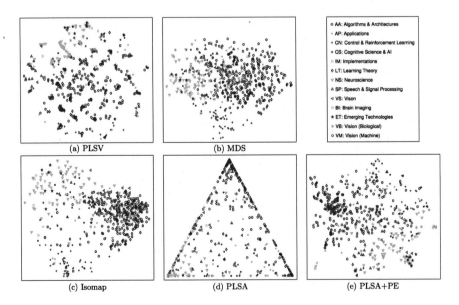

Fig. 4.2 Comparison of PLSV with different existing techniques [10]

ing an intuitive visual representation in which distances directly encode document similarities regarding to their topics.

In particular, PLSV computes the visualization coordinates of the input documents, $X = x_i{}_{i=1}^k$, based on a probabilistic topic model that captures the topic proportional of each document via Euclidean distance. Formally, the topic proportion of a document is defined as:

$$P(z|x_i, \Phi) = \frac{exp\left(-\frac{1}{2}||x_i - \phi_z||^2\right)}{\sum_{z'=1}^{Z} exp\left(-\frac{1}{2}||x_i - \phi_{z'}||^2\right)}$$

where $z \in Z$ is a latent topic and $\phi_z \in \Phi$ is its corresponding visualization coordinates. $||\cdot||$ is the L2-norm in the visualization space. Therefore, when a document x_i is close to a topic ϕ_z, the topic proportion is high, which indicates that the document has a higher probability to contain the content in the z-th topic. In this way, documents close to each other share a similar set of topics. The visualization results based on PLSV are shown in Fig. 4.2a and are comparison with the results visualized based on PLSA (Fig. 4.2d).

4.3 Conclusion

In this chapter, we introduce the techniques developed to illustrate the similarities between document files in a corpus. We have reviewed They produce visualizations in a similar form in which document files are represented as points on the display sized by importance and colored or shaped by semantics (such as topics). The screen distance between any pair of points indicates the similarities of the corresponding documents that are captured by various measures in different analysis models, following the rule of the closer, the more similar. Two major types of approaches are available: projection (i.e., dimension reduction)-based methods and semantic-oriented document visualizations; these visualizations are proposed for producing such a document overview that is developed based on different data and analysis models. In this section, we investigate these techniques and the corresponding document visualization systems.

References

1. Andrews, K., Kienreich, W., Sabol, V., Becker, J., Droschl, G., Kappe, F., Granitzer, M., Auer, P., Tochtermann, K.: The infosky visual explorer: exploiting hierarchical structure and document similarities. Inf. Vis. **1**(3/4), 166–181 (2002)
2. Balakrishnama, S., Ganapathiraju, A.: Linear Discriminant Analysis—A Brief Tutorial, vol. 18. Institute for Signal and Information Processing, Starkville (1998)
3. Balasubramanian, M., Schwartz, E.L.: The isomap algorithm and topological stability. Science **295**(5552), 7–7 (2002)

4. Blei, D.M., Ng, A.Y., Jordan, M.I.: Latent dirichlet allocation. J. Mach. Learn. Res. **3**, 993–1022 (2003)
5. Brandes, U., Pich, C.: Eigensolver methods for progressive multidimensional scaling of large data. In: Graph Drawing, pp. 42–53. Springer, Berlin (2007)
6. Gansner, E., Koren, Y., North, S.: Graph drawing by stress majorization. In: Graph Drawing, pp. 239–250 (2005)
7. Hinton, G.E., Roweis, S.T.: Stochastic neighbor embedding. In: Advances in Neural Information Processing Systems, pp. 833–840 (2002)
8. Hofmann, T.: Probabilistic latent semantic indexing. In: Proceedings of International ACM SIGIR Conference on Research and Development in Information Retrieval, pp. 50–57. ACM (1999)
9. Iwata, T., Saito, K., Ueda, N., Stromsten, S., Griffiths, T.D., Tenenbaum, J.B.: Parametric embedding for class visualization. Neural Comput. **19**(9), 2536–2556 (2007)
10. Iwata, T., Yamada, T., Ueda, N.: Probabilistic latent semantic visualization: topic model for visualizing documents. In: Proceedings of SIGKDD International Conference on Knowledge Discovery and Data Mining, pp. 363–371. ACM (2008)
11. Jolliffe, I.: Principal Component Analysis. Wiley Online Library (2002)
12. Jourdan, F., Melancon, G.: Multiscale hybrid mds. In: Proceedings of the IEEE Symposium on Information Visualization, pp. 388–393 (2004)
13. Kamada, T., Kawai, S.: An algorithm for drawing general undirected graphs. Inf. Process. Lett. **31**(1), 7–15 (1989)
14. Koren, Y., Carmel, L.: Visualization of labeled data using linear transformations. In: Proceedings of IEEE Symposium on Information Visualization, pp. 121–128 (2003)
15. Kruskal, J.B.: Multidimensional scaling by optimizing goodness of fit to a nonmetric hypothesis. Psychometrika **29**(1), 1–27 (1964)
16. Kullback, S., Leibler, R.A.: On information and sufficiency. Ann. Math. Stat. 79–86 (1951)
17. Van der Maaten, L., Hinton, G.: Visualizing data using t-SNE. J. Mach. Learn. Res. **9**(2579–2605), 85 (2008)
18. Morrison, A., Chalmers, M.: A pivot-based routine for improved parent-finding in hybrid MDS. Inf. Vis. **3**(2), 109–122 (2004)
19. Paulovich, F.V., Oliveira, M.C.F., Minghim, R.: The projection explorer: a flexible tool for projection-based multidimensional visualization. In: Computer Graphics and Image Processing, pp. 27–36. IEEE (2007)
20. Roweis, S.T., Saul, L.K.: Nonlinear dimensionality reduction by locally linear embedding. Science **290**(5500), 2323–2326 (2000)
21. Tenenbaum, J.B., De Silva, V., Langford, J.C.: A global geometric framework for nonlinear dimensionality reduction. Science **290**(5500), 2319–2323 (2000)
22. Williams, M., Munzner, T.: Steerable, progressive multidimensional scaling. In: Proceedings of the IEEE Symposium on Information Visualization, pp. 57–64 (2004)

Chapter 5
Visualizing Document Content

Abstract Text is primarily made of words and always meant to contain content for information delivery. Content analysis is the earliest established method of text analysis (Holsti et al., The handbook of social psychology, vol 2, pp 596–692, 1968 [55]). Although studied extensively and systematically by linguists, related disciplines are roughly divided into two categories, *structure* and *substance*, according to their subjects of study (Ansari, Dimensions in discourse: elementary to essentials. Xlibris Corporation, Bloomington, 2013 [9]). Structure is about the surface characteristics that are visible for a valid text, such as word co-occurrence, text reuse, and grammar structure. On the other hand, substance is the umbrella term for all information that needs to be inferred from text, such as fingerprinting, topics, and events. Various techniques have been proposed to analyze these aspects. In this chapter, we will briefly review these techniques and the corresponding visualization systems.

According to literary theory, written text situates syntactic symbols, such as words and punctuations, in a certain arrangement in order to produce a message. This arrangement of syntactic symbols is considered the message's **content** [71], rather than other physical factors, such as the appearance of the symbols or material on which it is written. Content analysis is the earliest established method of text analysis [55]. In the early years of content analysis, content was mainly viewed as the result of a communication process: "Who says what in which channel to whom and with what effect" [65]. Nowadays, this term is extended dramatically, covering from the communicative context of texts to their linguistic form. Although text content has been studied extensively and systematically by linguists, related disciplines are roughly divided into two categories, *structure* and *substance*, according to their subjects of study [9].

Structure is about all the defining characteristics that are visible at the surface of a valid text. Werlich [114] defined a text as

> an extended structure of syntactic unit such as words, groups, and clauses and textual units that is marked by both coherence among the elements and completion where as a non-text consists of random sequences of linguistic units such as sentences, paragraphs or sections in any temporal and/or spatial extension.

© Atlantis Press and the author(s) 2016   57
C. Nan and W. Cui, *Introduction to Text Visualization*, Atlantis Briefs
in Artificial Intelligence 1, DOI 10.2991/978-94-6239-186-4_5

All the units and the structure collectively contribute to the content of the text. At the structural level, researchers study how the syntactic units are organized with a strong logic of linguistic connections.

On the other hand, substance, which is a more subjective term, focuses on the information that can be inferred from the text, or on what remains after having "read" the text. The information is either intentionally delivered, such as the main idea of a document, or unintentionally delivered, such as the author of an anonymous blackmail note. Research in this category has largely been concerned with the ways people build up meaning in various contexts, with minimal attention paid to grammatical rules or structures. For example, discourse analysis aims to produce valid and trustworthy inferences via contextualized interpretations of messages produced in a communication process. Plagiarism detection is the process of detecting plagiarism instances within one or multiple documents by comparing particular patterns of language use, such as vocabulary, collocations, pronunciation, spelling, and grammar.

In this chapter, two sections are dedicated to analysis and visualization techniques for structure and substance, respectively. However, before these two sections, we decide to add one more for word visualization. There are two main reasons that we separate and put it first. One is that word is the most important syntactic unit for higher level analysis of structure and substance. It is part of text, but does not strictly belong to either structure or substance. The other reason is that bag-of-words is the fundamental model used in the text mining field. In this model, text content is treated as a set of words, disregarding grammar and even word order but keeping frequencies of occurrences. Since it is a very popular and successful model in mining, various visualizations have been specifically proposed for it. Therefore, before we introduce more sophisticated visualization solutions for structure and substance, it is necessary and meaningful to cover visualization techniques that are designed for the basic building block: word.

To summarize, this chapter mainly consists of three levels of visual analytics (from concrete to abstract): word, structure, and substance, which basically agrees with Justin Johnson's saying about text:

> not only what we say and how we say it, but also what we do not say which can be inferred from what we say.

5.1   "What We Say": Word

Besides words, it is clear that there are more things we can use during communications, such as tones and punctuations. However, we narrow our scope to words in this section due to its dominant position in content.

A word is the smallest language element that can deliver literal or practical meanings. As a consequence, the bag-of-words model is widely adopted to represent text contents. Intuitively, a bag of words contains the occurrences of words in the corresponding text regardless their position in it. The concept of "bag of words" is first mentioned by linguist Zellig Harris in 1954 [49]:

> ...for language is not merely a bag of words but a tool with particular properties which have been fashioned in the course of its use.

It is clear that Mr. Harris does not favor the idea of "bag of words". However after decades of research and testing, the bag-of-words model has proven successful in various research domains, such as content analysis, text mining, natural language processing, and information retrieval. It is generally considered a success of statistics: linguistic items with similar distributions have similar meanings. Based on this basic distributional hypothesis, researchers exploit the fast processing capability of computers and the massive text data available on the Internet to achieve good results:

> Every time we fire a phonetician/linguist, the performance of our system goes up.
> — Roger K. Moore, 1985.

Simple and powerful as it is, some information is clearly lost when using the bag-of-words model. First, each word may have multiple meanings that can only be understood as a part of different sentences. However, since the context of individual occurrences is ignored in the bag-of-words model, the meanings are treated equally. Second, the meaning of a whole sentence generally cannot be considered a summation of the meanings of its component words taken individually. Thus, the sophisticated information delivered by sentences are also completely lost in the bag-of-words representation.

Nonetheless, the bag-of-words model is still a popular and successful way to help people gain a quick overview about the text. In this section, we will look into visualizations that are specifically designed for the bag-of-words model.

5.1.1 Frequency

In the bag-of-words model, a valid text content is generally considered a set of words with occurrence frequencies in the corresponding text regardless of their positions in it (a bag of words). Thus, the data to be visualized for the bag-of-words model are essentially a set of [word, frequency] pairs, and a good visualization should deliver these two pieces of information effectively and intuitively.

In written language, words already have a well-accepted visualization form in written language, which are called *glyphs*. In typography, a glyph is a visual mark within a predefined set of marks intended to represent a character for the purpose of writing. For example, the letter *a* may appear differently using different font types, such as *Arial* and *Times New Roman*. However, all the appearances are valid visualizations of the letter *a*. In other words, glyphs are considered marks that collectively add up to visualize a word. Thus, we believe no visualizations can be more familiar, precise, or intuitive than directly showing the words (the data itself). For convenience, the term *word* may refer to the glyphs that visually represent the word or the semantic meaning behind it based on the term's context in the rest of this section.

Regarding the frequency information in the bag-of-words model, it is only natural to map it to other visual attributes of glyphs, such as font size, typeface, font weight, text color. As a matter of fact, several visual attributes have been adopted to encode additional information (not necessarily occurrences) of words in practice. For example, Amazon uses color intensity to tell users about how recently a tag is used. Del.icio.us uses red and blue to show if a tag is shared or not. However, among all the visual attributes, font size is most popular for representing a large range of numbers [12], such as occurrence frequencies. This group of visualizations that are composed of a set of words with size encoding one extra attribute, commonly the frequency value, are called *word clouds*. Word clouds are the dominant method for visualizing a bag of words, and thus, it is our focus in this section.

Background of Word Clouds

The concept of word clouds can be roughly traced back to Milgram's *psychological map of Pairs* [79]. In 1976, The social psychologist Stanley Milgram asked people to name landmarks in Paris, and drew a map with words with font size encoding the number of votes for the corresponding landmark (Fig. 5.1).

However, it was the photo sharing website Flickr that popularized the use of word clouds worldwide. When people upload photos to Flickr, tags are also attached to

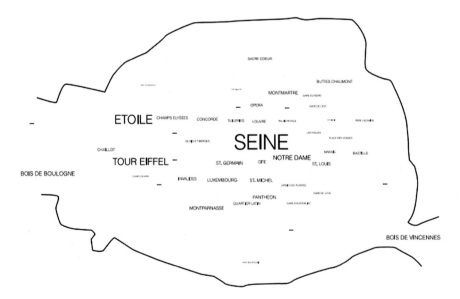

Fig. 5.1 Stanley Milgram's mental map of Pairs [79]. Each word is roughly placed at the location of the corresponding attraction of interest on the map

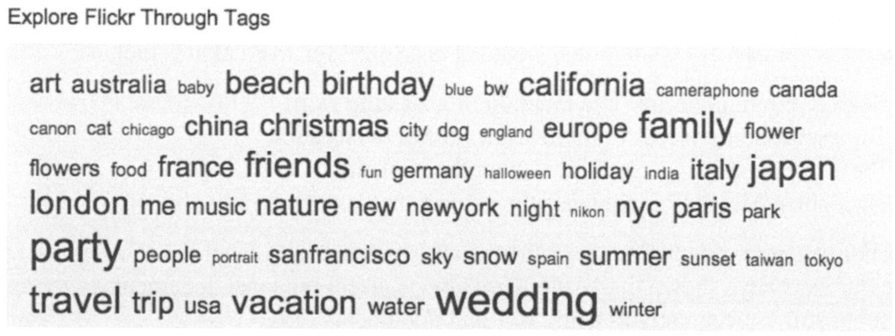

Explore Flickr Through Tags

art australia baby beach birthday blue bw california cameraphone canada canon cat chicago china christmas city dog england europe family flower flowers food france friends fun germany halloween holiday india italy japan london me music nature new newyork night nikon nyc paris park party people portrait sanfrancisco sky snow spain summer sunset taiwan tokyo travel trip usa vacation water wedding winter

Fig. 5.2 Example of a Flickr tag cloud (https://www.flickr.com/photos/tags)

label the photos. For example, a photo taken at a beach may have one or more tags, such as "beach", "sunset", "sea", or "scenery", based on the judgment of the person who uploads it. Flickr summarizes and displays some of the most popular tags using word clouds on its webpage,[1] which is probably why people used to call it *tag cloud* back then. With the help of tag clouds, users can glance and see what tags are hot and click to browse all the photos that carry the same tag (Fig. 5.2).

Nowadays people generally use "word cloud" and "tag cloud" interchangeably. Although visually the same, we still like to think they are slightly different in concept according to data sources: predefined meta-data or freeform text content. Tags in a tag cloud are semantically equal and equally meaningful. For example, tags in the tag clouds generated by Flickr represent different groups of photos, and are not necessary related to one another. By contract, words in a word cloud are often semantically related to one another, since they share the same content context. Thus, to best understand a word cloud, users may need to consider many words in the word cloud collectively. In addition, due to the roles of different words in the content, the words in a word cloud are also of different importance for making sense of the whole content. For example, the functional word *is* is clearly less useful than the noun *table* if they appear in the same word cloud.

Following the footsteps of Flickr, several other famous websites, such as Technorati and del.icio.us, also deploy word clouds to facilitate navigation. Word clouds have quickly become ubiquitous and abused on the Internet, no matter they are helpful or not. As a consequence, people rapidly grow tired of them. In fact, Flickr apologized for starting the craze about word clouds in 2006.

Just when word clouds are about to fade out on the Internet due to the abuse, researchers and software developers have found another arena, in which word clouds can be particularly useful: visualizing text contents.

There are reasons for word clouds being user friendly in general. Unlike other visualizations, word clouds need no additional visual artifacts and are self-explanatory. Users read the words and compare their sizes to know which one is more salient.

---

[1] https://www.flickr.com/photos/tags.

Simple as it is, word clouds are an excellent way to provide users with a high-level overview for casual explorations. Detailed tasks [89] for word clouds include:

- Search: Finding the size and location of a specific word.
- Browse: Finding words of interest for further exploration.
- Present: Forming a general idea about the content.
- Recognize/Match: Recognizing the subject that is described by the word cloud.

For example, one tedious task in text classification is to label the training data. Researchers have shown that using word clouds, users can label documents twice as fast but still as accurately as using full-text documents [93].

On the other hand, there are also some situations where word clouds are not appropriate. For example, people complain that word clouds cannot help make sense of or gain insight into a news article or a book [48]. Breaking text into word is like decomposing an electronic device into atoms. By checking the composing elements, although we can see that it is likely an electronic device, we certainly have no idea whether it is a television or an oven. Sometimes, word clouds may even mislead. For example, when we see a big word *nice* in the word cloud, we may consider it a positive impression on something. However, it is quite possible that the word *nice* in the content may have another word *not* co-occurring with it. The main reason is the context for each word is totally lost in word clouds. However, to be fair, we cannot completely blame word clouds for not being more meaningful. In fact, we believe that the issue is inherited from the bag-of-words model. Clearly, the context information is already lost at the data preparation process, so it is really impossible to recover at the visualization stage. If we just consider word clouds as a visualization technique to represent a bag of words, we would argue that there is no better way than word clouds. In the rest parts of this chapter, we will introduce more informative visualization techniques to represent text content. As far as word clouds, we need to be very careful about what tasks it is useful for.

Layout of Word Clouds

Generally, there are three ways to generate a word cloud:

- Sorted: Words are displayed as a sorted list based on dictionary order or frequency values.
- Random: Words are randomly placed in a 2D space.
- Cluster: Words are placed in a 2D space to reflect the cluster relationships between them via Euclidean distance.

Technically, cluster-based layout is not a visualization technique for a bag of words, since it requires cluster relationships between words that certainly are not contained by a pure bag of words. In fact, a cluster relationship is more like a type of structure information that indicates how words are organized. Therefore, we will mainly focus on the first two in this section, and introduce the third one in later sections.

Sorted-based layout is the oldest and easiest way to draw word clouds. It is commonly seen on websites like Flickr for navigation purposes. There are no sophisticated

Fig. 5.3 Examples of sorted-based tag clouds [90]: (*left*) based on alphabetical order and (*right*) based on frequency value

algorithms required to generated sorted-based word clouds. Words are simply displayed in a sequence with their font sizes equal to the relative importance or frequency values. Since there are only two pieces of information, i.e., word and frequency, for each item in a bag of words, we can only sort words based on alphabetic or based on frequency (Fig. 5.3). However, due to the irregular font sizes, wasted white space is inevitable between lines and words. All the unintentional white spaces are not aesthetically pleasing.

Unlike sorting-based methods that place all words in a sequence, random-based approaches freely place words on a plane and pack them together to generate a dense layout, which essentially falls into the category of geometric packing problems. Many algorithms have been designed, such as force-directed [32], seam-carving [115], and greedy [107], to remove the empty space between words. Intuitively, the first two algorithms start with a sparse, overlap-free layout of words, and then try to remove the empty space between words as much as possible. Different intuitions can be leveraged to achieve this goal. For example, Cui et al. [32] assume that there are attractive forces between words and use a force simulation to eliminate the white space between words. On the other hand, Wu et al. [115] construct an energy field based on the word distribution and remove the empty space that has lowest energy. The third category, greedy-based methods, is a little different. It directly adds words one-by-one to an empty canvas. Given a new word, the algorithm tests a set of candidate locations to place the word one-by-one, and checks if the word overlaps with the ones that have been added. Then, the algorithm adds the word to the canvas when a legit place is found, or rejects the word when all locations have been tested.

No matter which method is adopted, one key issue is efficiently detecting whether two word glyphs overlap, which essentially is a collision detection problem. One simple solution, which is also the most often adopted, is bounding box approximation (Fig. 5.4(left)). In this case, testing whether two words w_a and w_b overlap is equivalent to testing if their minimum bounding boxes $\Box_a$ and $\Box_b$ overlap, which can be efficiently obtained by checking the following value:

$$(\Box_a.L < \Box_b.R) \wedge (\Box_a.R > \Box_b.L) \wedge (\Box_a.T > \Box_b.B) \wedge (\Box_a.B < \Box_b.T),$$

Fig. 5.4 Two ways to pack words [32]: (*left*) treating each word as a rectangle and (*right*) treating each word as a complex shape

where edges of $\square_a$ and $\square_b$ are parallel to x-axis or y-axis, respectively. L, R, T, and B represent the left, right, top, and bottom bounds of the corresponding rectangle, respectively.

However, it still will not be really compact (Fig. 5.4(left)), since the word glyphs likely has ascent and descent empty spaces. Therefore, we need a more accurate and efficient way to detect the collision between word glyphs. Generally, there are two approaches, quadtree-based and mask-based, that are popular for achieving this goal.

A quadtree is a tree structure used to partition a 2d space by recursively subdividing it into four regions. When applying it to approximate the shape of a word, we first need to obtain the bounding box of the word. Then we recursively divide the bounding box into small four quadrants until one of the following conditions is met:

- there is no drawing of the word in the quadrant, which is considered clean; or
- the quadrant is full of drawing of the word, which is considered dirty; or
- the size of the quadrant has reached a threshold, which is also considered dirty.

Although building a quadtree for a word is expensive, the cost is reclaimed by an order of magnitude during the collision detection step. When testing for collisions between word w_a and w_b, we only need to recursively traverse down into the intersecting rectangles from the corresponding $tree_a$ and $tree_b$. The two words overlap if and only if two dirty leaves are found during the traversal. By pruning a lot of branches during the traversal, the results can be obtained very efficiently (Fig. 5.5).

For the other mask-based method, we also treat the drawings of words as images, each pixel is encoded with "1" and "0" to indicate whether it is occupied. Then we convert the matrix that is filled with ones and zeros to integers. By doing so, checking if two drawings becoming doing *bitwise-and* operation between integers, which can be efficiently performed in CPUs (Fig. 5.6).

Extending Word Clouds

Although intuitive, word clouds have limited capability of expression. Researchers have explored various paths to add additional values to it, such as supporting more interactions and encoding more information.

For example, **ManiWordle** [61] allows users to manually control the layout results produced by Wordle [107], a greedy algorithm automatically generates a static word

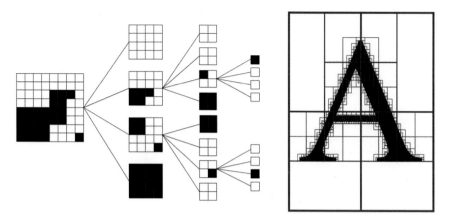

Fig. 5.5 Example of quadtree representation: (*left*) the construction of the quadtree for the shape example and (*right*) the quadtree result of the shape of letter *A*

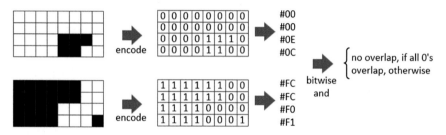

Fig. 5.6 Example of mask representation. Two drawings are converted to integers. Then *bitwise-and* operation is performed on the integers to check whether these drawings overlap

cloud in which words are randomly placed to achieve good compactness. In a typical random-based layout algorithm, users have no control over the final locations of any words in the generated layout. However, there are often scenarios when users have a preference for the final layout, such as putting two words side-by-side. ManiWordle is designed to accommodate this need. In a layout generated by Wordle, users may drag or rotate a word, ManiWordle will efficiently re-arrange the overlapped words and generate another version of compact word cloud accordingly (Fig. 5.7). Generally, their implementation also follows the basic idea of Wordle, such as collision detection, and spiral-based arrange strategy. Their efficient implementation makes sure the manual adjustments can be done in real-time. Another important issue during this interaction is animation. Since user interactions may trigger relocation of many words, animation is necessary to help users track these changes, such as location changes and angle changes.

RadCloud [15] is another technique targets at merging multiple word clouds into one. This technique is particular useful for comparison tasks. For example, different medias, such as news, blogs, and micro-blogs, may cover different aspects of the

Fig. 5.7 Example of ManiWordle [61]. (*Left*) The word *data* is rotated by user interaction; all the other words that overlap with the rotated *data* are pushed away to the closest legit locations to avoid overlapping. Empty space may appear during the process. (*Right*) All the words whose locations are not specified by users are re-arranged to generate a packed layout again

same social event. By comparing word clouds generated from individual sources, people may quickly grasp the difference between them without going through a tedious reading process. One intuitive idea is to generate word clouds independently for each source and display them side-by-side (Fig. 5.8(left)). However, finding the same and different words between these word clouds would be a nightmare. Instead, merging these clouds into one can help users quickly identify the shared words and unique words owned by individual word clouds. Figure 5.8(right) shows an example of RadCloud design. All words are placed inside the circle that consists of four colored arcs representing four different data sources. A word is initially placed at the center of the circle. Then it is moved towards different arcs based on its weight in the corresponding data sources. For example, the word *tag* appears in both green and orange data sources. Thus, two vectors are calculated accordingly and added together to decide the final location of the word *tag*, which is in the lower part of the circle. In addition, it is moved slightly to the left, since its size in the green source is bigger than that in the orange source. For words that are uniquely owned by a source, such as the word *word*, they are placed far from the circle center and close to the corresponding arc. The stacked bars chart placed under each word explicitly indicate how the corresponding word is shared by all sources.

Sometimes, word clouds are also placed on top of maps to expose text information that is tied to a specific location on a map. Thus, tag maps can be considered to be tag clouds grounded in real geographical spaces.

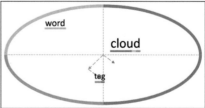

Fig. 5.8 (*Left*) Word clouds extracted from four text sources. (*Right*) The RadCloud representation that merges the four word clouds into one [15]

Fig. 5.9 Example of Tag Map [97]. (*Left*) Tag map and (*right*) tag cloud of the top 20 business directory searches, centered on South Manhattan

For example, **Tag maps** [57] are tag clouds overlaid on maps, in which the position of words is based upon real geographical space (Fig. 5.9). For an online map, zooming and panning are two basic and common operations to explore maps at different granularity. In this case, since the tags are constrained and tied to specific geographical locations, the main issue for this type of word clouds is to ensure word legibility during zooming and panning operations. During the interaction, words need to be selected based on their prominences and screen real estate in real time as part of the exploration process.

5.1.2 Frequency Trend

In a typical bag of words, all words and their associated frequency values are fixed. However, it is also easy to imagine that the bag of words can evolve as the data source change over time. During the evolution, new words may appear in the bag while some words may disappear. But, most likely, the frequencies of existing words change over time. Tracking the evolution can sometimes reveal interesting patterns and insights. One good example is **Google Ngram Viewer**, which is an online search engine that shows frequencies of any set of words or phrases using a yearly count in Google's text corpora between 1500 and 2008. This is a huge collection, 450 million words from millions of books, for users to freely search. For example, users can easily track and compare the popularities of Albert Einstein, Sherlock Holmes, and Frankenstein (Fig. 5.10). Simple as it is, the tool has been a huge help to socialists and historians in discovering cultural and historical patterns from the graphs of word

Fig. 5.10 Example of Google Ngram Viewer: the yearly counts of Albert Einstein, Sherlock Holmes, and Frankenstein in Google's text corpora between 1800 and 2000

usages. For more details about Google Ngram Viewer, please refer to the TED talk of Aiden [5].

Essentially, it is a classic multi time series visualization problem, in which a time series represents the frequency change of a word. William Payfair has studied this visualization by inventing multi-line charts in 1786 [43]. Figure 5.10 shows an example of such a visualization, which is so popular that everybody is used to it nowadays. However, there are limitations to applying line charts directly to a bag of words with changing frequency values. First, building connections between words and individual visual elements is less intuitive. Typically, words are either placed close to the corresponding lines (Fig. 5.10) or displayed as a legend next to the chart. However, neither of these ways is as user friendly as word clouds. Second, when there are many words in the bag, the line chart plot will become very cluttered and less appealing/clear. Visualization is about showing data aesthetically and informatively. Therefore, researchers have tried several alternatives to line charts for visualizing word frequency trends. We basically categorize existing attempts into three groups: small multiples, animation, and timeline.

Showing Frequency Trends with Small Multiples
Small multiples, also known as trellis chart, grid chart, panel chart, or lattice chart, are a visualization concept popularized by Edward Tufte [104], who described them as:

> Illustrations of postage-stamp size are indexed by category or a label, sequenced over time like the frames of a movie, or ordered by a quantitative variable not used in the single image itself.

Intuitively, small multiples use similar graphs or charts to show different partitions of a dataset. To enable efficient comparison, these charts are usually placed side-by-side and share the same measure, scale, size, and shape. If the sequence of charts follows a chronological order like comic books, they can be used to show changes. Therefore, it is easy to extend static word clouds to show frequency trends of individual words. For example, critical time points can be analyzed and extracted from the whole evolution. Then word clouds can be generated for all time points and displayed together as small multiples to see the trends (Fig. 5.11).

The advantage of this approach is simplicity. However, the readability of this visualization is rather low. In small multiples, users need to locate the same word in every word cloud to interpret its frequency changes. Thus, we cannot follow popular random-based algorithms, such as Wordle [107], to generate individual word clouds, since these algorithms make the locations of the same word in different word clouds unpredictable. Instead, new algorithms need to be designed to avoid the random-arrangement issue. The basic idea is that if some words are placed together in one layout, they are most likely placed together in another one.

Cui et al. [32] are the first to propose a context-preserving word cloud generation approach to achieve this goal. Figure 5.12 illustrates the pipeline of their algorithm. First, words are extracted from all text content in disregard their appearances on the time line (Fig. 5.12a). One big sparse word cloud is generated from these words accordingly, which largely defines the relative positions between words at all small multiples (Fig. 5.12b). To generate the word cloud for a specific time point, irrelevant words are temporally removed from the big sparse word cloud, which results in an even sparser word cloud (Fig. 5.12c1). Then, a triangle control mesh is constructed to connect all words in the filtered, overlap-free word cloud (Fig. 5.12d1). The mesh is the key to removing empty spaces and keeping relative positions stable. Abstractive forces are exerted on the mesh edges to make sure words move close to one another. During the move process, collision detection is applied to avoid word overlapping. In addition, the mesh are consistently checked for planarity status. As long as the mesh

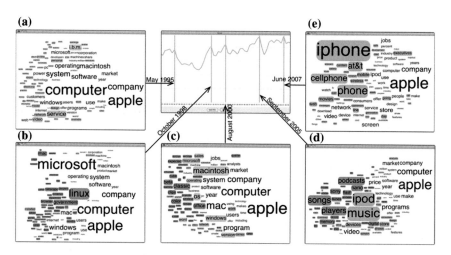

Fig. 5.11 Example of showing trends with small multiples [32]. (*Top center*) The *curve* shows the content fluctuation extracted from a collection of news articles related to Apple Inc. The x-axis encodes the time and the y-axis encodes the significance of the word clouds at individual time points. Five word clouds, **a–e**, are created using the algorithm presented in [32] for five selected time points where high significance values are observed

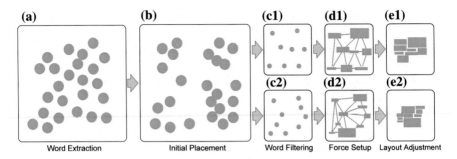

Fig. 5.12 The pipeline for context-preserving word cloud generation [32]

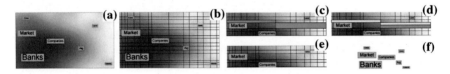

Fig. 5.13 The pipeline for seam-carving based word cloud generation [115]

stays planar, we believe the relative position of words are kept (Fig. 5.12e1). When generating the word cloud for another time point, the process is repeated (Fig. 5.12c2–e2). Since the relative positions of words are kept during the space-removal process, the layout stability can be preserved between different word clouds.

As another example, Wu et al. [115] use seam-carving techniques to achieve this goal. Originally, seam-carving [11] is a content-aware image resizing algorithm. It first estimates an energy field based on the content of the image, and then repeatedly selects a seam of the lowest energy crossing the image (left-right or top-bottom) to duplicate or remove. In this case of word clouds, the empty space between words is considered to have the lowest energy for removal. Figure 5.13 shows the basic process. First, a sparse word cloud is obtained, just like the one obtained in Fig. 5.12c1. The red background indicates the Gaussian importance field based on the word arrangement (Fig. 5.13a). Then, the layout is divided into grids according to the bounding boxes of the words (Fig. 5.13b). An optimal seam (marked in blue) that crosses the layout (left to right in this case) is selected (Fig. 5.13c). To prevent distorting of the layout, the seam is pruned to have a uniform width (yellow seam in Fig. 5.13d). Figure 5.13e shows the word cloud layout after the seam is carved. The compacted word cloud result is obtained when there are no more seams to be carved (Fig. 5.13f).

Showing Frequency Trends with Animation

The second way is animation, which has been used to show changes for many years [91]. When the frequency of a word changes, it is natural to see its size change accordingly as the size is used to encode the frequency of the word. Although animation may work well with individual words, things become complicated when animating a word cloud. First, since word clouds are usually compact, directly changing word sizes without moving the words will cause them to overlap. Thus, the key issue in animating a word cloud is resolving the overlap during size animation. One possible solution is using key-frames. We can generate a sequence of overlapping-free word clouds representing several key points on the time line. For example, we may consider Fig. 5.11a–e as keyframes and make an animation by interpolating the word size and locations in them. However, although there is no overlapping in each key-frame, words still may overlap during the interpolation process, which may cause some readability and aesthetic issues. To solve this issue and achieve better animation effect, rigid body dynamics [22] can be used to ensure word clouds are overlap-free during whole animation. Rigid body dynamics do not assume keyframes. Instead, each word is regarded as a rigid body that can collide with others during the size changing process and moves based on two momentums, linear and angular. The dynamics start with an initial configuration, and then a numerical solver is used to track the change of the state of each word during the animation process. To ensure a smooth and pleasing visualization, several constraints are proposed and can be optionally integrated into the dynamics.

- Contacting: used to avoid word overlapping by assigning reaction forces to contacting words.
- Boundary: used to specify the legit space to arrange words and help ensure the compactness. This is achieved by viewing the boundary as a fixed wall. All words that touch the wall are totally rebounded.
- Orientation: used to ensure all words have the same angle. This is achieved by specifying a uniform rotation matrix to all words and ignoring their angular momentums.
- Position: used to fix the location of specific words by directly setting their positions and ignoring their linear momentums.

Showing Frequency Trends with a Timeline
In the third category, time is generally encoded as an axis in the display. As a matter of fact, the multi-line representation invented by William Playfair [43] falls into this category. However, many variations have been designed to alleviate the limitations of traditional line charts (Fig. 5.14).

The first alternative is generally called stacked graphs, a type of graph that show time series data by using stacked layers, originally invented by William Playfair [85]. The primary advantage of stacked graphs is allowing users to see individual time series in the context of the aggregation of all data. Only recently have versions that can scale up to large number to time series been created. **ThemeRiver** [50] is probably the first attempt to enhance the power of stacked graphs to visualize a bag of words/tags that evolves over time. A layer in the ThemeRiver representation represents the frequency trend of a specific term (or "theme") in a news feed. The novelty of ThemeRiver is two-fold. First, the frequency values at different time points are interpolated to generate a smooth stripe, in contrast to long polygons in a

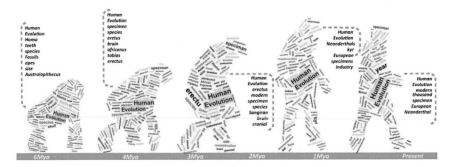

Fig. 5.14 Example of rigid body dynamics result [22]: a morphable word cloud shows human evolution. The word frequencies changed, and the words are arranged inside a shape that morphs from Australopithecine to Homo sapiens. In particular, the word *Human Evolution* is fixed during animation using the position constraint

traditional stack graph, so that users can easily track the change. Second, layers are not stacked on the x-axis in ThemeRiver, but rather in a symmetrical shape, which also greatly improves the aesthetics of the visualization. As a follow-up to ThemeRiver, **Streamgraph** [16] further discusses the aesthetic limitations of stacked graphs, and proposes a new layout algorithm that aims to reduce the layer wiggles (Fig. 5.15).

Compared with traditional multi-line chart, ThemeRiver-like visualizations have better scalability, and allow users to intuitively see a layout with the context of layer aggregation. However, the labeling issue still exists in stack graphs. Since each word is converted into a stripe, the label information is lost in this visual encoding. For a layer that is thick enough, the corresponding word can be placed inside the layers. However, for those layers that do not have enough space, interactions are generally adopted to help users build the connections between words and stripes.

Parallel Tag Cloud (PTC) [26] is another way to visualize the word frequency change directly. In a PTC visualization, the x-axis also represents time. Words that are extracted from the documents at each time slot are arranged vertically. The words are sorted alphabetically with sizes encoding the word frequencies at individual slots. Thus, the same words may not be aligned horizontally, such as the highlighted

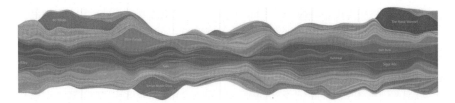

Fig. 5.15 Streamgraph visualization of a person's music listening history in the last.fm service [16]. The x-axis encodes time and each layer represents an artist. The varying width of a layer indicates how many times the person has listened to songs by a given artist over time

word *patent* in Fig. 5.16. However, users can interact with the system to explicitly see the word connected to see their trends. For those words that are not currently focused, small lines segments are placed to on both sides of the word to indicate the connections, which is used to avoid extensive visual clutter.

A third popular example is called **SparkClouds** [67]. In this work, the authors integrate sparklines into word clouds to show the trends for words. A sparkline [105] is generally a line chart that piggybacks on another visual element to indicate the trend in some measurement of the piggybacked item in a simple and highly condensed way. It is typically small and drawn without axes or coordinates (Fig. 5.17). Lee et al. [67] test different alternative arrangements of sparklines and evaluate their performance in six tasks related to trend comparison or interpolation. Their results have shown that this simple technique is intuitive and better than other techniques, such as multi-line charts and stacked bar charts.

Fig. 5.16 Parallel Tag Cloud (PTC) visualization [26]. X-axis represents time. When users hover over a word, its appearances at different time points are linked together to show trends

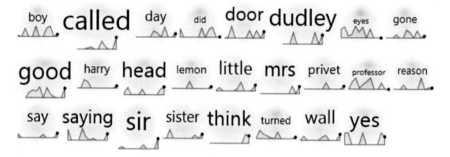

Fig. 5.17 SparkClouds visualization [67]. The word sizes may either encode the total frequencies over the entire time period or the frequencies at a specific time point. The sparklines beneath words illustrate the overall frequency trends of corresponding words over time

Which Method is the Best?

Here we briefly discuss the pros and cons of the aforementioned methods, since none of them is perfect for all scenarios. Showing trends with small multiples is a natural extension of static word clouds. It is storyboard style visualization that is familiar to everyone and easy to understand. However, tracking words between different word clouds is the biggest pain for this type of methods. Although techniques have been proposed to alleviate this issue by ensuring word relative relations stable across different word clouds, users still find it not easy. Another issue is the scalability. To make sure words are legible, each word cloud cannot be too small. Thus, small multiples of word clouds are also limited by screen real estate. For example, in a typical display with a resolution of 1920 by 1080, only five or ten word clouds can be displayed side by side to ensure effective comparisons.

For animation-based approaches, they also inherit the drawback of animation, confusing people in certain scenarios. Previous perception research [86] has suggested that people can only track a subset of up to 5 objects simultaneously in order to distinguish a change in a target from that in a distractor. A good animation should ensure a clear and clean story. Thus, we can expect that if all words are changing their sizes and locations chaotically, users can easily get confused and lose focus. Although animations are generally fun and engaging, finding frequency trend or patterns using animation often requires repeatedly playing, pausing, or rewinding the animation. Robertson et al. [91] formally compare the performances of animated and static visualizations for trend tracking tasks. Their results show that the core value of animation is for presentation. However, users still slightly prefer using static visualizations for analysis tasks.

The third category is the most popular and commonly seen in many visual analytics systems [30, 38, 74], which are not limited to visualizing time-varying word clouds, but time series data in general. First, it uses statistic representation, which avoids the drawback of animation. Second, it provides strong visual cues to help users track changes, avoiding the drawback of small multiples. However, since one dimension (x-axis) is used to encode time, only one dimension can be exploit to encode additional information. For example, words can no longer be arranged free in a 2D space, like what typical word clouds do. In certain scenarios [32, 115], where spatial relationships between words have meanings such as semantic distance between words, timeline-based approaches are certainly less expressive then the other two.

5.2   "How We Say": Structure

The word "text" originates from the famous statement in Quintilian's book on speeches:

> "after you have chosen your words, they must be weaved together into a fine and delicate fabric," – with the Latin for fabric being *textum*.

Thus, a text is always expected to have a well-defined structure. To deliver sophisticated messages, different words must be arranged appropriately and correspond to one another in a manner conditioned by logic and grammar.

There are various ways a word or syntactic unit can correspond to another. For example, one of the most studied relationships is known as *co-occurrence*, which may refer to words appearing in the same document, in the same sentence, following the same word, etc. According to the definition of co-occurrence, the distance and the order between words are irrelevant. Considering the aforementioned bag-of-words model, if all words are extracted from a document, the corresponding word cloud will be a visualization for document-wise co-occurring relationships. However, for more sophisticated relationships or structures, such as concatenation and grammar relationships, word clouds would certainly be incapable.

In this section, we formally review and summarize existing work that visualize different types of structure in text.

5.2.1 Co-occurrence Relationships

Co-occurrence is the most common relationship studied in text mining and natural language processing applications. They are linguistic assumptions behind co-occurrence: if two words co-occur, they are semantically related, neglecting the distance and order between one another. This assumption makes using the bag-of-words model to represent the content in a document reasonable. Document-wise co-occurrence has been widely used in many applications, such as content analysis, text mining, construction of thesauruses and ontologies [76].

Although document-wise co-recurrence is the most adopted in existing literature, co-occurrence can be defined at other granularities. For example, content sometimes may dramatically change within a document. Thus, we may also investigate co-occurrence in smaller units, such as paragraphs or sentences. Besides defining co-occurrence with such linguistic units, we can consider the words in a fixed-width window surrounding the target word. Since the window width is a parameter, different values may apply. In general practice, 5, 20, and 100–200 word windows approximate phrase, sentence, and paragraph co-occurrence, respectively. In certain scenarios, co-occurrence is even considered within a syntactic structure. This grammar-aware co-occurrence can help cluster terms more delicately, such as finding all nouns that follow the verb "drink": water, milk, bear, etc.

Statistics of co-occurrence may be represented in two ways. For example, let us compute the co-occurrence number of the word *to* and *be* in the context of the sentence "to be or not to be". The simplest way to check if the sentence contains both words in disregard of their appearance number. In this case, the co-occurrence value $C_{to \cdot be} = C_{be \cdot to} = 1$. Alternatively, we may also consider the two *to*'s and two *be*'s are different, adding their co-occurrence together, which give us $C_{to \cdot be} = C_{be \cdot to} = 4$. In the later case, people imply that the contexts that contain repeated target words are different from the contexts that do not. Given a set of contexts, e.g., all sentences in a

Fig. 5.18 Example of a co-occurrence between words

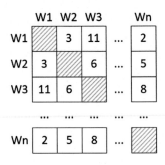

documents, a co-occurrence matrix can be then obtained by aggregating all pair-wise co-occurrence values together (Fig. 5.18).

Co-occurrence matrices tell people whether two specific words co-occur or how strongly they co-occur. More importantly, they help people find correlations between word pairs and/or similarities of meaning among/within word patterns. The underlying assumption of statistical semantics was established by Harris [49] and popularized by Firth [41]:

A word is characterized by the company it keeps.

To be more specific, two words that tend to have similar co-occurrence patterns, tend to be positioned closer together in semantic space and resemble each other in meaning. Mathematically, each column or row in a co-occurrence matrix is considered the feature vector defined by the co-occurrence pattern between the corresponding word and all other words. Therefore, if two words "drink" and "eat" exist in the same context, we would not be surprised to see them have similar vector values since they resemble each other in meaning.

As mentioned before, word clouds can be used to visualize co-occurrence relationships. For example, a word cloud consisting of words from a document can be considered a document-wise co-occurrence visualization. In particular, Vuillemot [108] proposes a system, **POSvis**, to help users search specific words and explore the surrounding words (Fig. 5.19).

However, it is also natural to use graphs or links to explicitly describe co-occurrence relationships. For example, users may have one or more focused keywords and show the other terms that co-occur with the focused keywords explicitly (Fig. 5.20(left)). When users have no particular focused keywords, it is also helpful to show a general graph that depicts strongly co-occurred word pairs extracted from content (Fig. 5.20(right)).

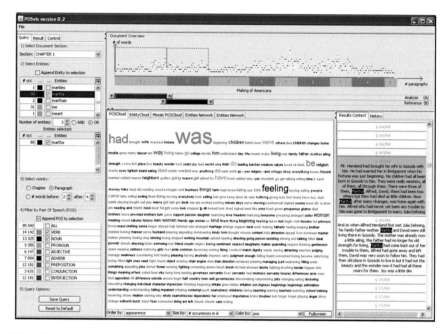

Fig. 5.19 POSvis visualizing Gertrude Stein's *The Making of Americans* [108]. The focused word *Martha* is selected by users. The co-occurred verbs and nouns are displayed in the centered main window as a word cloud

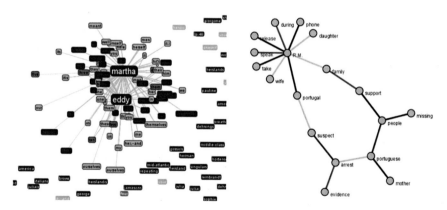

Fig. 5.20 Examples of using links to represent co-occurrence relationships. (*left*) Terms that co-occur with either "martha" or "eddy" are connecting to these two words explicitly with links [108]. (*right*) Keyword pairs that have co-occurrences passing a given threshold are connected together as a general graph [100]

5.2.2   *Concordance Relationships*

While co-occurrence relationships are mainly exploited to infer the similarity between different words, concatenation concerns more about the ways in which the words are connected within a consecutive sequence. Formally, Sinclair [96] describes this relationship as *concordance* in 1991:

> A concordance is a collection of the occurrences of a word-form, each in its own textual environment. In its simplest form it is an index. Each word-form is indexed and a reference is given to the place of occurrence in a text.

Originally, a concordance is a manually prepared list of words found in a text along with references to their locations in the text, which is a tedious work requiring a lot of human effort to build such an index. For example, as the first recorded concordance in history, *Correctio Biblie* was built by Hugh of St-Cher with the help of hundreds of Dominican monks in Paris in the year 1230'.

To demonstrate the concept of concordance, Sinclair [96] gives an example of one-word context concordance of a short text: "The cat sat on the mat." (Fig. 5.21). All word occurrences are placed in the middle column of the table. The previous and next words (if existing) are placed in the first and third columns, respectively.

The concordance relationships are originally recorded to provide scholars with the information about where words were used in a closed set of text and a source of insight and illumination for a wider readership. With the capability of generating concordances in the blink of an eye, computers can support a wider range of applications than its paper counterpart.

To begin with, it can help with better author profiling. Idiolect already shows that linguistic variation not only exist at the word level, but also at the phrase level or sentence level. For example, some may say "Use your feelings, Obi-Wan, and find him you will," while the others prefer saying "Obi-Wan, use your feelings, and you will find him". Thus, concordances can preserve more clues/information for classification models to better identifying authorships. In this case, the bag-of-words model will certainly fail, since the data are identical.

Another major application of concordance relationships is word prediction, which can be used in any tasks where we have to identify words in noisy/ambiguous input, such as speech recognition, handwriting recognition, spelling correction, or machine translation. The problem is generally considered as: in a document collection, given some preceding words, what is the most likely word to follow? The likelihood is

Fig. 5.21 One-word context
of all words extracted from
"The cat sat on the mat." [96]

the -------→	cat -------→	cat
the -------→	mat -------→	
sat -------→	on -------→	the
cat -------→	sat -------→	on
-------→	the -------→	cat
on -------→	the -------→	mat

generally captured with simple statistical method based on Markov assumption: the probability of a word depends only on the probability of a limited history.

The keyword in context (KWIC) visualization (Fig. 5.22) is the first designed and most commonly used tool for concordance analysis. The visualization consists of a list of text fragments with the searched keyword aligned vertically in the view. Thus, users can easily compare or inspect linguistic properties of the words before or after the keyword, which is usually highlighted with a different color.

However, the traditional KWIC visualization generally shows concordances in great detail, and is not intuitive for statistical information, such as variation and repetition in context. Recently, **Word Tree** [112] and **Double Tree** [33] have been proposed to address this issue. Word Tree visualizes all contexts of a keyword using a tree layout. Considering the occurrences of adjacent words, Word Tree expands tree branches to encode the recurring phrase patterns. Compared with Word Tree that only shows one side of context, Double Tree shows both sides (Fig. 5.23).

Unlike traditional KWIC-like visualizations that focus on the surrounding words of a specific words, **Phrase Nets** [106] focuses on visualizing specific concordance relationships, such as is- and of-relationships. Specifically, Phrase Nets builds a graph, in which nodes represent words and edges represent a user specific concordance relationships (Fig. 5.24).

5.2.3 Grammar Structure

Text, although seeming linearly developed, consists of a complicated network constructed by grammar. Since concordance related models only concern the ways in which words are connected into a sequence, they are often criticized for lacking

```
1     emed quite natural);  but when the  Rabbit  actually TOOK A WATCH OUT OF ITS WA
2     t a thousand times as large as the  Rabbit,  and had no   reason to be afraid of
3      `No, they're not,' said the White  Rabbit,  `and that's the  queerest thing ab
4     ging for apples, indeed!' said the  Rabbit  angrily.  `Here!  Come and help me
5     che!'      Alice watched the White  Rabbit  as he fumbled over the list,  feeli
6      `Did you say "What a pity!"?' the  Rabbit  asked.          `No, I didn't,' said Al
7      `She boxed the Queen's ears--' the  Rabbit  began.  Alice gave a  little scream
8     d the King.        On this the White  Rabbit  blew three blasts on the trumpet, a
9     ess,' said the King; and the White  Rabbit   blew three blasts on the trumpet,
10    n't opened it yet,' said the White  Rabbit,  `but it seems  to be a letter, wri
11    ask help of any one; so, when the  Rabbit  came near her, she began, in a low
12    e afraid of it.      Presently the  Rabbit  came up to the door, and tried to o
13    he stairs.  Alice knew it was the  Rabbit  coming to look for her, and  she tr
14    she stopped hastily, for the White  Rabbit  cried out, `Silence in  the court!'
15    ight brass plate with the name `W.  RABBIT'  engraved upon it.  She went in wi
16    jury.        `Not yet, not yet!' the  Rabbit  hastily interrupted.  `There's  a g
17                          Down the  Rabbit-Hole            Alice was beginning to
18    n time to see it pop  down a large  rabbit-hole under the hedge.        In anoth
19    she was to get out again.      The  rabbit-hole went straight on like a tunnel
20    most wish I hadn't gone down that  rabbit-hole--and yet--and yet--it's rather
```

Fig. 5.22 KWIC visualization for the word *rabbit* in *Alice in Wonderland*

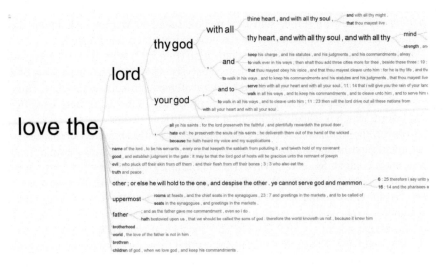

Fig. 5.23 Word Tree visualization [112] of the King James Bible showing the contexts to the right of *love the*

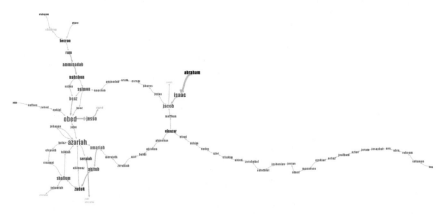

Fig. 5.24 Phrase Nets visualization [106] of *X begat Y* relationships in Bible, revealing a network of family structures

grammatical dependency information. Thus, in this section, let us move one step further into complicated relationships and consider grammar structure in text.

So, what is grammar? The term *grammar* is derived from the Greek word *grammatikē*, where *gram* means something written, and *tikē* derives from technē and meant art. Hence, grammar is the art of writing, which has a long history tracing back to ancient Greece and Rome. After thousands of years and going through several stages of development, such as prescriptive grammar, non-structural descriptive grammar, structural descriptive grammar, and generative grammar, the term nowadays is generally defined as the set of structural rules governing the composition of clauses, phrases, and words in any given natural languages.

Although various grammar frameworks have been proposed, there are usually two ways to describe sentence structure in natural languages, constituency and dependency. Although proven to be strongly equivalent [45], they focus on different aspects of grammar.

Traced back to the ancient Stoics [77], constituency grammar emphasizes part-whole relationships between words and greater units, and now becomes the basis of grammar model in computer science. The basic structure for constituency grammar analysis is subject-predicate. A typical sentence is divided into two phrases: subject, i.e., noun phrase, and predicate, i.e., verb phrase. The phrases are then further subdivided into more one-to-one or one-to-more correspondence. Thus constituency relationships can be visualized as a tree structure (Fig. 5.25(left)). Every leaf in the tree corresponds to a word in the sentence. And a parent node is a bigger or more complex syntactic unit by combining its child nodes. Therefore, for each word in the sentence, there are one or more nodes containing it.

Dependency grammar [51, 94], on the other hand, is a relatively new class of modern linguistic theories that focuses on explaining any agreements, case assignments, or semantic relationships between words. Compared with constituency grammar, it emphasizes head-dependent relationships between individual words, and there are no phrases or clauses in the structure. All units in the dependency grammars are words. Each relationship involves exactly two words, i.e., a head and a dependent. In general, the head determines the behavior of the pair, while the dependent is the modifier or complement. Thus, the dependency relationships between words are usually visualized as curved arrows placed on top of the sentence (Fig. 5.25(top right)). Each arrow points from a head to its dependent. However, the hierarchy of the dependency relationships is unclear in the arc visualizations. Thus, trees (Fig. 5.25(bottom

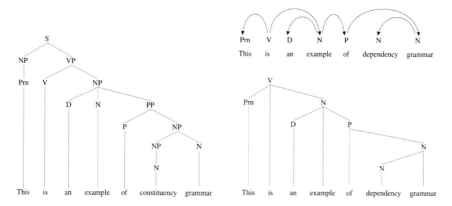

Fig. 5.25 Examples of grammar visualizations [27]: (*left*) a tree visualization of constituency grammar structures, (*top right*) a arrow visualization of dependency grammar structures, and (*bottom right*) a tree visualization of dependency grammar structures

right)) are also often used to help researchers see the dependency relationships and the hierarchy simultaneously.

5.2.4   Repetition Relationships

Repetition is another interesting and common pattern in content, which serves different purposes and inspires further investigation into interpretations of the text. In a narrative or speech, for example, authors may use repetition to emphasize themes or ideas worth paying close attention to. Repetition may also occur between two or multiple texts. One typical example is plagiarism, which has gain much attention recently due to the availability of large, online document archives. On the other hand, the repetition can also appear in different ways. In many cases, the repetition may be exact or near exact by using the same words or phrases, such as direct quotes. In other cases, the recurrences may occur at a more subtle level, such as themes, motifs, and metaphors.

The most obvious and common repeated elements are words and phrases. Readers can easily spot repeated words in a document simply by reading it. However, recurrences of function words like "the" or "is" are trivial and meaningless. For effective understanding of text content, choosing the correct words and analyzing their repetition patterns in the content can help users generate high-level interpretations. A good visualization for repetition patterns should simultaneously show the total number of occurrences and the gaps between occurrences. For example, arc diagrams [111] are proposed to represent complex repetition patterns in string data (Fig. 5.26). Intuitively, identical subsequences in the whole sequences are connected by thick semicircular arcs. The thickness of the arcs is equal to the length of the repeated unit. The height of the arc is thus proportional to the distances between the two units. Since only consecutive pairs of identical subsequences are connected with arcs, users can easily follow the path of arcs and understand the distribution of a specific repeated subsequence.

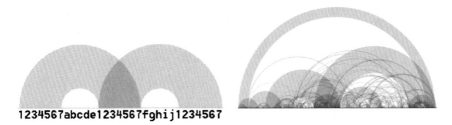

Fig. 5.26 Examples of arc diagrams [111]: (*left*) two arcs connect the three occurrences of a subsequence and (*right*) the arc diagram visualization of Beethoven's *Für Elise*

As another example, **FeatureLens** [37] focuses on frequent item sets of n-grams in a text collection. Figure 5.27 shows a screenshot of FeatureLens visualizing *The Making of Americans*. The left panel shows a list of precomputed n-grams for users to explore. Users can select one or more from the list (three in this screenshot), the views in the middle present the occurrences of these n-grams in individual speeches. A rectangle represents a speech, and the height of the rectangle encodes the number of paragraphs in the corresponding section. Each colored line in a rectangle indicates that the paragraph at the corresponding location contains the word. Darker lines suggest more occurrences of the corresponding words. Thus, users can compare the occurrences patterns of different features of interest. For example, users can easily identify the paragraphs where selected features occur together.

However, in many scenarios, the recurrences may not always happen at the copy-paste level. Similar ideas or concepts can be evoked or repeated using different words, which is impossible to capture by finding repeated words or phrases. Thus, a higher level of abstraction is required to capture conceptual similarity. To quantitatively capture the similarity between two fragments of text, many conceptual similarity algorithms have been proposed, such as Leximancer [98], Latent Semantic Analysis [64], and Latent Dirichlet Allocation [14]. For example, Picapica [1], a text reuse search engine, defines at least seven types of plagiarism, from complete plagiarism to translation plagiarism. Based on the search engine, Riehmann et al. [88] build a visual analysis tool to help users effectively assess and verify alleged cases of plagiarism.

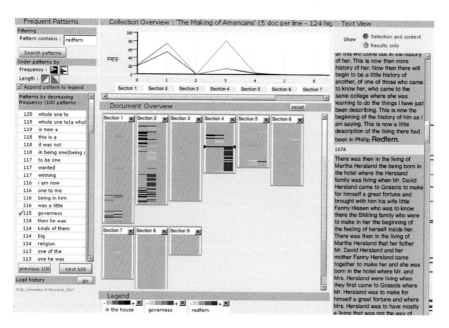

Fig. 5.27 FeatureLens visualization [37] showing three n-grams: *in the house*, *governess*, and *redfern*

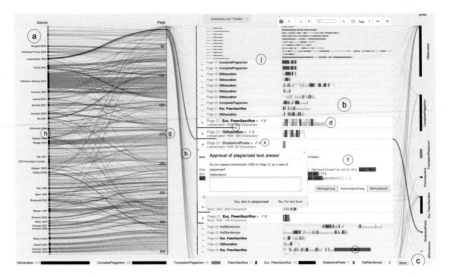

Fig. 5.28   Screenshot of the visual analysis tool built by Riehmann et al. [88]

Figure 5.28 shows a screenshot of their system. An overview of the distribution of alleged spots in the document is displayed in the left panel, along with their lengths, types, and relationships with the source documents. In the right view, glyphs are used to emphasize the relationships between the alleged fragments and the source text, which aims to help users quickly form a decision without looking at the actual text.

5.3   "What Can Be Inferred": Substance

In the two previous sections, we focus on dissecting the surface of text, which is essentially a set of expressions made of words. However, expressions are meant to deliver information. Some delivered information is explicit, while some other part of the information is implicit. Understanding the information is essential for many analysis tasks. Thus, in this section, we move beyond the surface of text and dig deeper into the semantic aspect of text.

5.3.1   Fingerprint

Linguists believe that every person has their own distinct way to use language, namely idiolect, which can be understood through distinctive choices in text. Thus, a category of methods called *linguistic fingerprinting* is developed to capture linguistic impressions from a text, just like a signature, to represent the text and identify the authors.

Linguistic fingerprinting is originally used for authorship. Since Jan Svartvik published *The Evans Statements: A Case For Forensic Linguistics* [102] in 1968, lawyers and courts have been repeatedly demonstrated its success in cases of disputed authorship of suicide notes, anonymous letters, confession statements, etc.

Recently, the rapid growth of the Internet has made it the largest public repository of anonymous information, such as emails [35] and forum messages [2, 69], and provided an unprecedented opportunity and challenge to authorship analysis. Various techniques have been proposed, which generally focus on two aspects, namely, style feature extraction and classification techniques.

Zheng et al. [123] summarize four categories of writing style features that can be extracted from a text for authorship identification, namely, lexical, syntactic, structural, and content-specific features. For example, lexical features include word counts, words per sentence, paragraph lengths, vocabulary richness, word distribution, etc. Syntax and structural features refer to the patterns or rules used to construct sentences and paragraphs, such as punctuations, stop words, and the use of greetings. Content-specific features are defined as keywords that are important for specific contexts. All these features are captured at the surface level of text, which is discussed in previous sections. However, the meaning the content is not longer important to linguistic fingerprinting. Instead, combined with machine learning techniques, such as support vector machines, neural networks, and decision trees, these features have shown great potential and accuracy in determining authorships of content [2, 35, 69].

One common concept in these approaches is *document fingerprint*, which aims to generate a compact, comparable description for each document. This concept is first proposed by Manber [75]. In this work, a fingerprint is defined as a collection of integers that encode some key content in a document. Each integer, also called *minutia*, is generated by selecting a representative subsequence from the text and applying a hashing-like function to it. All the minutiae are combined together and indexed to support quick query and identification of similar documents. The fingerprint concept is then extended and evolved in many approaches [52, 95], but there are two basic considerations that may greatly influence the accuracy of these approaches: granularity and resolution [54]. The size of text to generate a minutia is considered the fingerprint granularity, while the number of minutiae is considered the resolution.

Researchers have devoted great effort to creating visualization tools [3, 59, 82] to facilitate fingerprint analysis. In these approaches, style features, such as word frequency, average sentence length, and number of verbs, are extracted and visually integrated together to provide users with a high-level signature for the document. For example, Keim and Oelke [59] introduce **literature fingerprinting**, a pixel-based representation of text features. In this visualization, the document is segmented into blocks of different granularities, such as sections and chapters, and features are computed for each block, encoded as a colored pixel. The pixels are packed and arranged based on the document blocks to collectively provide a compact and scalable visualization of the document (Fig. 5.29). Their technique is then further extended to

Fig. 5.29 Literature fingerprinting visualizations [59] of Jack London's *The Iron Heel* and *Children of the Frost*. The *color* is used to indicate the vocabulary richness

help users analyze why particular paragraphs and sentences are difficult to read and understand [83].

Literature fingerprinting allows users to define different types of features at different levels, such as average word length, average number sentence length, and vocabulary richness. The computed feature values are shown in a heatmap (Fig. 5.29), in which each pixel represents a text block and its color indicates the values of one of the selected features whose range is represented by a legend at the side of the visualization. Although very simple, the compact visualization is deliberately designed as an image or a signature that helps users quickly compare to find (dis)similarities between multiple text content. Following the same design, Oelke et al. [82] introduce a fingerprint matrix to help understand a dynamic network extracted from the text.

Following a similar idea, Jankowska et al. [58] introduce the **document signature** which summarizes a document based on character n-grams. The character n-grams are consecutive sequences of n characters extracted from the given text. For example, we can extract four distinctive character 5-grams from the text "the␣book": "the␣b", "he␣bo", "e␣boo", and "␣book". Character n-grams have been widely used to handle text classification problems, such as language recognition [20], and authorship analysis [56]. In this work, Jankowska et al. [58] analyze and visualize the difference of n-gram patterns between two or multiple documents. For a pair of documents, namely a base document and a target document, a rectangle visualization, called a relative signature, is constructed to represent the difference between the most common n-grams between these documents. The n-grams are represented by horizontal stripes. The stripe colors indicate the difference of frequency patterns of the corresponding n-gram between these two documents. Intuitively, the white stripe means that the n-gram has similar or identical frequency patterns between these two document. The blue stripes at the top of each rectangle indicate that these n-grams only appear in the target document. The red-yellow scale indicate the frequency of the corresponding n-gram appear more frequently in the base document then in the target document (Fig. 5.30).

From the above examples, we can see that one of the major applications of the above fingerprint-based technique is to help with document comparison. Besides

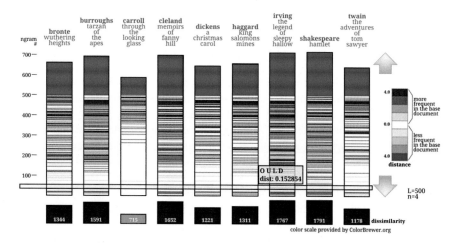

Fig. 5.30 The relative signatures of nine books with L. Carroll's *Alice's Adventures in Wonderland* as the base document [58]. It is clear that the signature of Carroll's *Through the Looking Glass* is much lighter and shorter than those of the other books, which indicates that the book is much more similar to the base book than the other books are. In addition, users can select an n-gram ("OULD" in this case) to compare its frequencies in different book

fingerprinting, several other techniques are also developed for the similar purpose. For example, Rohrer et al. [92] introduce a glyph design to summarize the text data by illustrating the features via shapes, which was used to study Shakespeare's works. Many visual analysis systems are also developed for plagiarism detection [87, 88] or authorship identification [3], or showing the change of language [53].

5.3.2 Topics

With the unprecedented growth of text data accessible on the Internet, people now more and more likely to run into tasks to digest of a large collection of documents. In this scenario, individual documents are no longer important. Instead, people need to step back from individual documents, construct an overview of the content in the corpus, and look at the larger pattern among all documents, which traditionally can only be obtained by going through a tedious reading and note-taking process.

To assist effective understanding of document collections, one intuitive way is to perform clustering on the documents [4, 17, 42, 72, 81, 118, 119]. Clustering of documents can automatically organize documents in a meaningful structure (either hierarchical or partitioning) based on their content to help users build a coherent conceptual overview of a document corpus [84]. On one hand, clustering has been widely studied as a fundamental aspect of machine learning, and many general clustering techniques, such as k-means, have been applied to document clustering. On the other hand, more specific clustering techniques have been proposed taking advantage of the characteristics of documents. Examples include latent semantic indexing [36], non-negative matrix factorization [119], and concept factorization [17]. These methods generally transform document features from the noisy and sparse term space to a concise latent space, which in turn, to help improve the clustering quality. Highly related to document clustering, topic modeling techniques have been developed to help people discover and summarize categories and patterns. In this section, we focus on the topic modeling and visualization techniques.

So what exactly does the term *topic* mean in the context of topic modeling? Intuitively, a topic is a short description of a group of semantically-related documents. In the domain of topic modeling, a topic is often represented by a set of words that frequently appear in documents but less frequently in other documents. The idea is also based on the distributional hypothesis of words. For example, "Jordan" and "basketball" may appear more often in documents about Michael Jordan, while "Jackson" and "music" may appear more often in articles about Michael Jackson. However, "Michael" may have no significant difference between both document sets. Thus, "Michael" is not an appropriate distinguishing topic term for either of the document sets. After years of development, a topic is now formally defined a probability distribution of words in a vocabulary, which aims to indicate the probability of each individual word appearing in a document on a specific topic [99].

Topic modeling is an umbrella term for machine learning techniques used to find topics in textural data. Its history can be traced back to the early 90s, when Deerwester et al. [36] proposed *latent semantic analysis* for clustering semantically-related words for natural language processing research. However, it is the Latent Dirichlet Allocation (LDA) model, proposed by Blei et al. [14] in 2003, that popularizes topic modeling in practice. The basic intuition of LDA is that a document usually cannot be neatly categorized into one single topic, but can relate to a set of topics. In other words, a document is a mixture of several topics with different weights.

The goal of LDA and many related topic modeling techniques is to identify these topics and quantify the proportions of these topics for every document. To achieve this goal, these methods consider this problem the other way around and assume a generative model for a text corpus. To begin with, we assume all the topics in the text corpus are known and the corpus is empty. To generate a document in the corpus, we first randomly choose a distribution over topics. Then, we repeatedly choose a topic based on the topic distribution. From each picked topic, we randomly pick a word based on the corresponding word distribution.

However, the reality is the opposite. We have the text corpus ready but unknown topics to be decided. Therefore, what a LDA-like topic modeling method tries to do is to estimate the parameters that make the model have a high likelihood to generate a corpus similar to the given corpus via the aforementioned generative process. The process of estimation is the key and differs various topic modeling approaches. For example, the original LDA model [14] is based on a process called variational inference. Gibbs sampling [47] is also very popular technique to discover hidden topics in a large text corpus. An excellent survey about probabilistic topic models can be found at [34].

The output of a typical topic modeling method is usually complicated. First, a set of topics is estimated from the corpus, and each topic consists of ranked probabilities of words. Similarly, each document is also summarized as the ranked probabilities of topics. Then, the topic modeling method can further cluster documents based on these distributions to identify representative documents for each topic. Although the text corpus is summarized to a certain extent, the complicated output is still challenging to evaluate and interpret. Thus, various topic visualization techniques have been developed specifically to help users with this task.

To start with, Chuang et al. [25] propose a matrix-style visualization called Termite to help users organize and evaluate term distributions associated with topics generated by LDA model (Fig. 5.31). In this work, the authors define a saliency measure and a seriation method to arrange the terms in the matrix to emphasize clustering structures in the dataset.

Many other visualizations are built upon projection techniques [18, 19, 24, 46, 68, 70]. For example, TopicNets [46] generates a document-topic graph by using an LDA model [10]. Then, the topics are represented as nodes that are projected onto a 2D surface via a multidimensional scaling technique [28]. Once the topic nodes are fixed, a force-directed layout algorithm [44] is applied to the document-topic graph and fix the positions for each document node (Fig. 5.32).

UTOPIAN [24] is another visualization that highly integrates with a Nonnegative Matrix Factorization (NMF) topic model. It also uses a projection-based visualization to help users explore the topic space. In many cases, the results generated by a topic model are not perfect. Since NMF is a deterministic model that can generate

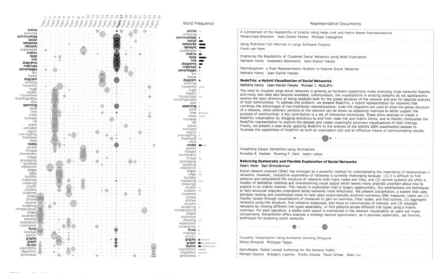

Fig. 5.31 Termite visualization [25] of a LDA result. The term-topic matrix is displayed to the *left*. When a topic is selected in the matrix (Topic 17 in this case), the related terms are highlighted and their frequency distribution is displayed in the middle. In addition, the most representative documents are listed to the *right*

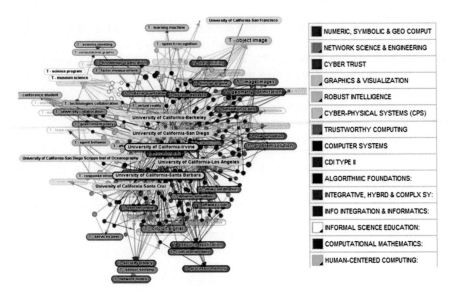

▣	NUMERIC, SYMBOLIC & GEO COMPUT
▣	NETWORK SCIENCE & ENGINEERING
▣	CYBER TRUST
◣	GRAPHICS & VISUALIZATION
▣	ROBUST INTELLIGENCE
◣	CYBER-PHYSICAL SYSTEMS (CPS)
▣	TRUSTWORTHY COMPUTING
▣	COMPUTER SYSTEMS
▣	CDI TYPE II
▣	ALGORITHMIC FOUNDATIONS:
▣	INTEGRATIVE, HYBRD & COMPLX SY:
▣	INFO INTEGRATION & INFORMATICS:
▢	INFORMAL SCIENCE EDUCATION:
▣	COMPUTATIONAL MATHEMATICS:
◣	HUMAN-CENTERED COMPUTING:

Fig. 5.32 TopicNets visualization [46] of the document-topic graph generated from CS-related NFS grants that are awarded to University of California campuses. The topic nodes are labeled with the top two frequent terms and *colored* based their fields. The document nodes are displayed as *small circles filled with the corresponding colors*

a consistent result from multiple runs, UTOPIAN allows users to progressively edit and refine the topic results by interactively changing the algorithm settings. In the visualization, UTOPIAN uses colored dots to represent documents in different topics, exploiting a technique called t-distributed stochastic neighborhood embedding (t-SNE) to generate the 2D layout to reflects the pairwise similarities between documents (Fig. 5.33).

5.3.3 Topic Evolutions

Many text corpus have time stamps associated to individual document in it, such as news articles, forum posts, email achieves, and research papers. For these text corpora, the temporal patterns of the topics are the key in many analysis tasks. Thus, related mining and visualization techniques have been proposed specifically for this goal.

One simple solution is to first ignore the time stamps of documents and generate topics as usual with the aforementioned topic modeling techniques [40, 113]. Once the topics are generated, the documents that related to each topic can be split based on their time stamps to help users to understand how the topics are distributed on the time dimension. However, for streaming data, it is impossible for a topic model

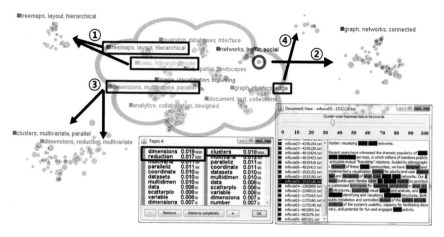

Fig. 5.33 UTOPIAN visualization [24] of the topics in the collection of InfoVis papers (1995–2010) and VAST (2006–2010). The topic result can be refined via a rich set of interactions, such as topic merging/splitting and topic inducing

to know all data in advance. Thus, the data are generally processed in batches. For example, evolutionary clustering algorithms [21, 23, 29, 31, 80, 117] aim to find a clustering result for the new batch of the streaming data, such that the result is coherent with those of the arrived batches and appropriately reflect the structure in the new batch. Topic detection and tracking (TDT) [6, 121] is designed to detect new topics or find documents that belong to already detected topics from a document stream. The LDA model is also extended to handle streaming data [13, 109, 110].

Similar to the bag-of-words model, the attribute of time also dramatically changes the visualizations of topics. Most of current dynamic topic visualizations are based on stacked graphs [29, 31, 101, 113, 116]. For example, TIARA [113] first applies LDA model to a text corpus to generate topics, and then uses a probability threshold to assign one or more topics to each document in it. In the TIARA visualization, each topic is visualized as a layer in a stacked graph. The x-axis encodes time, and the thickness of a layer at a time point is proportional to the number of documents that belong to the topic and have the corresponding time stamps (Fig. 5.34).

Cui et al. [29] further design TextFlow, a river-based visualization, to convey relationships between evolving topics (Fig. 5.35). In contrast to stacked graphs that show individual stripes evolving independently, TextFlow allows stripes to split or merge during evolution, which can intuitively deliver the complex splitting/merging patterns between topics. In addition, relationships between words are also visualized using a thread-weaving metaphor. Users can selectively show words as threads overlaid on top of the stripes. If two or more words co-occur during a specific time span in the same topic, the corresponding threads also intertwine to generate a weaving pattern in the visualization to convey such information.

The expressive visualization cannot be achieved without proper support of the topic modeling and visualization techniques. On the backend of TextFlow, an

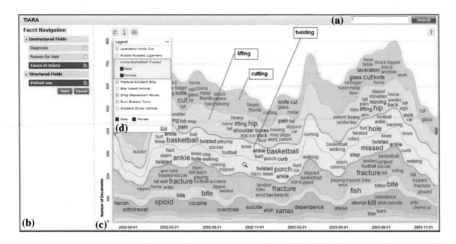

Fig. 5.34 TIARA visualization [113] of over 23,000 clinical records (2002–2003). Each layer represents a topic discovered by the LDA model. The *overlaid labels* represents the top key words in the documents of the corresponding time and topic

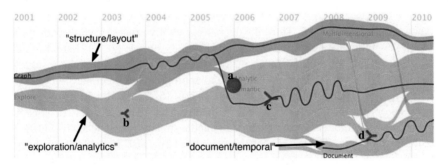

Fig. 5.35 TextFlow visualization [29] of selected topics in InfoVis papers (2001–2010). Three main topics are visualized as three stripes that have splitting/merging relationships. Several keywords are visualized as threads that weave together to convey their co-occurrence relationships

incremental hierarchical Dirichlet process model [103] is adopted to learn the splitting/merging patterns between topics. In addition, a set of critical events and keywords are ranked and indexed to represent the content at different time points. Finally, a three-level directed acyclic graph layout algorithm is design to integrate and present the mining results in a coherent and visually pleasing visualization.

Similar to TextFlow, RoseRiver [31] follows the same river metaphor to depict the splitting/merging patterns between evolving topics. However, in this work, the authors focus on the scalability issue of datasets. In contrast to TextFlow that builds flat topic structures, RoseRiver builds a hierarchical topic tree for each time point and makes sure the topic trees are coherent at neighboring time points. This approach has two advantages over the TextFlow implementation. First, different users may have different interests or focuses for the same corpus. RoseRiver can provide a "one

(a)

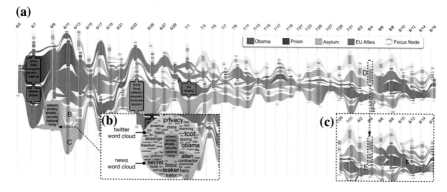

Fig. 5.36 RoseRiver visualization [31] of news topics related to the Prism scandal (June 5–Aug 16, 2013). Four major topics are highlighted with *red*, *purple*, *green*, and *blue*, respectively. If users are interested in one topic at a specific time point (marked with *D*), they can split and examine smaller topics that contained by it (see **c**)

shoe fits all" solution to them. For topics that are not of interest, users can choose a coarse granularity and hide details to avoid distractions. In contrast to RoseRiver, TextFlow may need to generate different flat topic structures to meet specific needs of different users. Second, since the visualization is supported by topic hierarchies, users can freely refine and alter the topic granularities during the exploration process by interacting with the system (Fig. 5.36).

5.3.4 Event

Cognition studies [60, 120] have shown that people are often used to perceive and make sense of a continuously observed activity by segmenting it into a sequence of related discrete events. However, the definition of *event* is controversial. For example, the Merriam-Webster dictionary generally defines it as "something that happens," "a noteworthy happening," or "a social occasion or activity". Zacks and Tversky [120] define it as "a segment of time at a given location that is perceived by an observer to have a beginning and an end." Kim [60] considers an event a structure consisting of three elements: object, a property, and time (or a temporal interval). In the topic detection and tracking field, an event is something that has a specific topic, time, and location. Although expressed differently, these definitions all similarly reflect people's intuition about events, but certainly do not exhaust the common conception of events. Ironically, many researchers also admit that none of these features are essential for an event.

In this section, we do not intend to contribute to this debate about the accurate definition of event. Instead, we look at events from the perspective of text analysis, and try to understand how people use events to efficiently make sense of a dynamic text corpus. When following a narrative or an evolving topic, readers often encounter plots or occurrences that impact or change the course of development. These plots or occurrences, which may present themselves at multiple levels of granularity [63], are generally critical to helping readers comprehend the logic and reasoning behind the development. Thus, they are often considered the primary type of events to detect and analyze in many text visualization approaches.

As one of the first attempts, topic detection and tracking [7, 8] was a DARPA-sponsored initiative investigating techniques to find and follow new events in news streams back in 1996. The basic idea is straightforward. Their algorithm first extracts features from a newly arrived article to form a query. Then, the query is compared against earlier processed queries. If no earlier queries are triggered by exceeding a certain threshold, the new article is marked as containing a new event. Otherwise, the new article is marked as containing an earlier event. There are many follow-up methods based on the same idea. An excellent survey on topic detection and tracking approaches can be found in [6]. For example, Mei and Zhai [78] use a language model to extract and analyze the temporal structure of events. Luo et al. [73] assume that news articles are generated based on real-word events. In the assumption, once a worthy real-world event happens, news agencies will generate a set of articles that have closely-related contents in a continuous time period while the event draws continuous attention. Thus, they believe that a big temporal-locality cluster of news articles that have highly related contents may indicate and represent a real-world event that motivate the documents in it. The authors propose a two-stage temporal-locality clustering algorithm to extract these events, and further cluster events into groups to represent long-term stories.

In many other approaches, researchers will first apply topic models to organize a text corpus into topics that perpendicular to the time dimension, then segment each topic into a sequence of events to extract critical changes in the topic. For example, Zhao and Mitra [122] first extract a hierarchy of topics from the content of social text streams, such as emails. Then, given the sequence of emails within a topic, they also construct a hierarchy of temporal segments to represent events at different levels of granularity. In addition, the author information of these emails are also exploited to help users group and explore these events with the context of social communities. Dou et al. [39] define an event in a text stream as "an occurrence causing change in the volume of text data that discusses the associated topic at a specific time". Based on this definition, the authors apply the accumulative sum control chart [66] to locate volume bursts, which are treated as events, in individual topic streams. Intuitive, the algorithm tracks a accumulative sum of volume at each timespan. Once the sum exceeds the predefined threshold, the corresponding timespan will be considered as an event. In each event, name entities including people and location are also extracted to help users explore relationships between events.

Once the events are successfully extracted from text streams, their visualizations are rather unanimous in different approaches. For example, EventRiver [73] visualize an event (a temporal-local cluster of articles) as a bubble in a river of time (Fig. 5.37). The varying width and the size of the bubble indicate the attention it draws over time

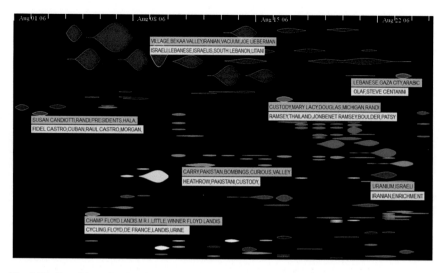

Fig. 5.37 EvenRiver visualization [73] of CNN news (29,211 articles from Aug 1 to Aug 24, 2006)

and in total, respectively. A novel layout algorithm is also used to pack event bubbles of similar contents together to form a long-term story.

CloudLines [62] defines events as "time-stamped data points that arrive in data streams at non-equidistant time intervals." For example, the authors apply this concept to news streams and consider a news article mentioning a specific person as an event. Then, they group events by person and visualize each group as a ribbon on top of the time-axis (Fig. 5.38). The density and width of a ribbon can help users quickly identify important event episodes for further detailed analysis.

Similar to CloudLine, LeadLine [39] first extracts topics from a dynamic text corpus and visualizes them as ribbons with width encoding the topic strength. Based on this familiar visual metaphor, the authors use the accumulative sum control chart

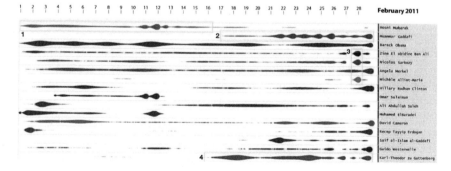

Fig. 5.38 CloudLines visualization [62] of events related to 16 politicians who appeared in news articles in Feb 2011

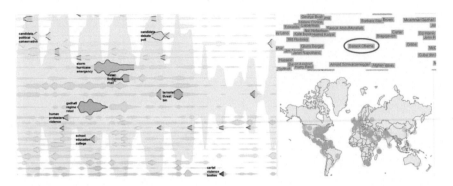

Fig. 5.39 CloudLines visualization [62] of events related to 16 politicians who appeared in news articles in Feb 2011

to locate time spans in each topic, and highlight them with colors and contour lines as critical events in the corresponding topic. In addition to the detected events, the authors also extract name entities of people and location from the events, and visualize them in two juxtaposed views to help users make sense of these events (Fig. 5.39).

5.4   Summary of the Chapter

In this chapter, we focus on techniques that are widely used for analyzing text content. From simple and concrete to complicate and abstract, we progressively review the basic content analysis techniques in three steps.

The first step is words, the fundamental building element of text. We focus on the bag-of-words model and related visualization techniques in this step. Although the basic idea is simple, the bag-of-words model has been proven effective and successful in many application domains. Due to its simplicity, the related visualizations, such as Wordle [107], have also become popular and been accepted by more and more people on the Internet.

The second step is structures. In the bag-of-words model, words are treated equally and independently. However, text content is much more than the meanings of individual words combined. In this second step, we take a step further to explore relationships between words, and look into the methods that people use to understand how words are constructed to build text content. There are various ways we can examine text structures. For example, grammar structure is obvious one way to understand word relationships. However, grammar structures are mainly exploited to help computers make sense of texts, and not very helpful when shown in visualizations. On the other hand, other statistical relationships are more helpful, such as word co-occurrences, word concordances, and text reuses. The statistics of these relationships can help us not only to see what words are used, but also understand how they are used. In addition, more contexts of individual words are provided and help us make sense of contents more easily.

The third step is substance. In this step, we leave the surface of text behind and dig deeper into the information that can be inferred from texts. First, we look at the linguistic fingerprinting. In fingerprinting analysis, the actual text content is no longer important. Instead, researchers focus on developing techniques to capture linguistic impressions from text expressions, just like signatures, to represent the text, differentiate linguistic styles, and identify the authors. We then introduce techniques that aim to provide high-level summarizations of large text corpora. Specifically, we focus on machine learning models and visualizations that help us discover and understand topics and events in text corpora. These techniques ignore individual documents and minute details. Instead, they extract and organize the most important and essential information, and then use it to construct a concise overview of a text corpus, which traditionally can only obtained through a tedious and intensive close reading process.

Content analysis is a large area, and we just briefly cover the most common and popular aspects in these sections. Many technical details are skipped, and many rich possibilities remain open to explore and catalog. As we enter the era of big data, it is more and more challenging to perform traditional content analysis: reading. Thus, we are never more convinced that computed-aided content analysis is the future, and hope readers can draw inspiration from our discussions and encourage further study of this area.

References

1. Picapica. www.picapica.org (2014). [Online; accessed Jan. 2016]
2. Abbasi, A., Chen, H.: Applying authorship analysis to extremist-group web forum messages. Intelligent Systems, IEEE **20**(5), 67–75 (2005)
3. Abbasi, A., Chen, H.: Visualizing authorship for identification. In: Intelligence and Security Informatics, pp. 60–71. Springer (2006)
4. Aggarwal, C.C., Zhai, C.: A survey of text clustering algorithms. In: Mining Text Data, pp. 77–128. Springer (2012)
5. Aiden, E.L., Michel, J.B.: What we learned from 5 million books. https://www.ted.com/talks/what_we_learned_from_5_million_books (2011). [Online; accessed Jan. 2016]
6. Allan, J.: Topic detection and tracking: event-based information organization, vol. 12. Springer Science & Business Media (2012)
7. Allan, J., Carbonell, J.G., Doddington, G., Yamron, J., Yang, Y.: Topic detection and tracking pilot study final report (1998)
8. Allan, J., Papka, R., Lavrenko, V.: On-line new event detection and tracking. In: Proceedings of the 21st annual international ACM SIGIR conference on Research and development in information retrieval, pp. 37–45. ACM (1998)
9. Ansari, T.: Dimensions in Discourse: Elementary to Essentials. Xlibris Corporation (2013)
10. Asuncion, A., Welling, M., Smyth, P., Teh, Y.W.: On smoothing and inference for topic models. In: Proceedings of the Twenty-Fifth Conference on Uncertainty in Artificial Intelligence, pp. 27–34. AUAI Press (2009)
11. Avidan, S., Shamir, A.: Seam carving for content-aware image resizing. In: ACM Transactions on graphics (TOG), vol. 26, p. 10. ACM (2007)

12. Bateman, S., Gutwin, C., Nacenta, M.: Seeing things in the clouds: the effect of visual features on tag cloud selections. Proceedings of the nineteenth ACM conference on Hypertext and hypermedia **4250**, 193–202 (2008). doi:10.1145/1379092.1379130. URL http://portal.acm.org/citation.cfm?id=1379130

13. Blei, D.M., Lafferty, J.D.: Dynamic topic models. In: Proceedings of the 23rd international conference on Machine learning, pp. 113–120. ACM (2006)

14. Blei, D.M., Ng, A.Y., Jordan, M.I.: Latent dirichlet allocation. the. Journal of machine Learning research **3**, 993–1022 (2003)

15. Burch, M., Lohmann, S., Beck, F., Rodriguez, N., Di Silvestro, L., Weiskopf, D.: Radcloud: Visualizing multiple texts with merged word clouds. In: Information Visualisation (IV), 2014 18th International Conference on, pp. 108–113. IEEE (2014)

16. Byron, L., Wattenberg, M.: Stacked graphs-geometry & aesthetics. Visualization and Computer Graphics, IEEE Transactions on **14**(6), 1245–1252 (2008)

17. Cai, D., He, X., Han, J.: Locally consistent concept factorization for document clustering. Knowledge and Data Engineering, IEEE Transactions on **23**(6), 902–913 (2011)

18. Cao, N., Gotz, D., Sun, J., Lin, Y.R., Qu, H.: SolarMap: Multifaceted Visual Analytics for Topic Exploration. 2011 IEEE 11th International Conference on Data Mining pp. 101–110 (2011). doi:10.1109/ICDM.2011.135. URL http://ieeexplore.ieee.org/lpdocs/epic03/wrapper.htm?arnumber=6137214

19. Cao, N., Sun, J., Lin, Y.R., Gotz, D., Liu, S., Qu, H.: FacetAtlas: Multifaceted visualization for rich text corpora. IEEE Transactions on Visualization and Computer Graphics **16**(6), 1172–1181 (2010). doi:10.1109/TVCG.2010.154

20. Cavnar, W.B., Trenkle, J.M., et al.: N-gram-based text categorization. Ann Arbor MI **48113**(2), 161–175 (1994)

21. Chakrabarti, D., Kumar, R., Tomkins, A.: Evolutionary clustering. In: Proceedings of the 12th ACM SIGKDD international conference on Knowledge discovery and data mining, pp. 554–560. ACM (2006)

22. Chi, M.T., Lin, S.S., Chen, S.Y., Lin, C.H., Lee, T.Y.: Morphable word clouds for time-varying text data visualization. Visualization and Computer Graphics, IEEE Transactions on **21**(12), 1415–1426 (2015)

23. Chi, Y., Song, X., Zhou, D., Hino, K., Tseng, B.L.: Evolutionary spectral clustering by incorporating temporal smoothness. In: Proceedings of the 13th ACM SIGKDD international conference on Knowledge discovery and data mining, pp. 153–162. ACM (2007)

24. Choo, J., Lee, C., Reddy, C.K., Park, H.: UTOPIAN: User-driven topic modeling based on interactive nonnegative matrix factorization. IEEE Transactions on Visualization and Computer Graphics **19**(12), 1992–2001 (2013). doi:10.1109/TVCG.2013.212

25. Chuang, J., Manning, C.D., Heer, J.: Termite: Visualization techniques for assessing textual topic models. In: Proceedings of the International Working Conference on Advanced Visual Interfaces, pp. 74–77. ACM (2012)

26. Collins, C., Viegas, F.B., Wattenberg, M.: Parallel tag clouds to explore and analyze faceted text corpora. In: Visual Analytics Science and Technology, 2009. VAST 2009. IEEE Symposium on, pp. 91–98. IEEE (2009)

27. Covington, M.A.: A fundamental algorithm for dependency parsing. In: Proceedings of the 39th annual ACM southeast conference, pp. 95–102. Citeseer (2001)

28. Cox, T.F., Cox, M.A.: Multidimensional scaling. CRC press (2000)

29. Cui, W., Liu, S., Tan, L., Shi, C., Song, Y., Gao, Z., Tong, X., Qu, H.: Textflow: Towards better understanding of evolving topics in text. IEEE Transactions on Visualization and Computer Graphics **17**(12), 2412–2421 (2011). doi:10.1109/TVCG.2011.239

30. Cui, W., Liu, S., Tan, L., Shi, C., Song, Y., Gao, Z.J., Qu, H., Tong, X.: Textflow: Towards better understanding of evolving topics in text. Visualization and Computer Graphics, IEEE Transactions on **17**(12), 2412–2421 (2011)

31. Cui, W., Liu, S., Wu, Z., Wei, H.: How Hierarchical Topics Evolve in Large Text Corpora. IEEE Transactions on Visualization and Computer Graphics **20**(12), 2281–2290 (2014). doi:10.1109/TVCG.2014.2346433. URL http://research.microsoft.com/en-us/um/people/weiweicu/images/roseriver.pdf

32. Cui, W., Wu, Y., Liu, S., Wei, F., Zhou, M., Qu, H.: Context-preserving, dynamic word cloud visualization. IEEE Computer Graphics and Applications **30**(6), 42–53 (2010). doi:10.1109/MCG.2010.102

33. Culy, C., Lyding, V.: Double tree: an advanced kwic visualization for expert users. In: Information Visualisation (IV), 2010 14th International Conference, pp. 98–103. IEEE (2010)

34. Daud, A., Li, J., Zhou, L., Muhammad, F.: Knowledge discovery through directed probabilistic topic models: a survey. Frontiers of computer science in China **4**(2), 280–301 (2010)

35. De Vel, O., Anderson, A., Corney, M., Mohay, G.: Mining e-mail content for author identification forensics. ACM Sigmod Record **30**(4), 55–64 (2001)

36. Deerwester, S., Dumais, S.T., Furnas, G.W., Landauer, T.K., Harshman, R.: Indexing by latent semantic analysis. Journal of the American society for information science **41**(6), 391 (1990)

37. Don, A., Zheleva, E., Gregory, M., Tarkan, S., Auvil, L., Clement, T., Shneiderman, B., Plaisant, C.: Discovering interesting usage patterns in text collections: integrating text mining with visualization. Main pp. 213–221 (2007). doi:10.1145/1321440.1321473. URL http://portal.acm.org/citation.cfm?id=1321473

38. Dörk, M., Gruen, D., Williamson, C., Carpendale, S.: A visual backchannel for large-scale events. Visualization and Computer Graphics, IEEE Transactions on **16**(6), 1129–1138 (2010)

39. Dou, W., Wang, X., Skau, D., Ribarsky, W., Zhou, M.X.: LeadLine: Interactive visual analysis of text data through event identification and exploration. Visual Analytics Science and Technology (VAST), 2012 IEEE Conference on pp. 93–102 (2012). doi:10.1109/VAST.2012.6400485. URL http://ieeexplore.ieee.org/xpl/articleDetails.jsp?arnumber=6400485

40. Dou, W., Yu, L., Wang, X., Ma, Z., Ribarsky, W.: Hierarchicaltopics: Visually exploring large text collections using topic hierarchies. Visualization and Computer Graphics, IEEE Transactions on **19**(12), 2002–2011 (2013)

41. Firth, J.R.: A synopsis of linguistic theory, 1930-1955 (1957)

42. Forsati, R., Mahdavi, M., Shamsfard, M., Meybodi, M.R.: Efficient stochastic algorithms for document clustering. Information Sciences **220**, 269–291 (2013)

43. Friendly, M., Denis, D.J.: Milestones in the history of thematic cartography, statistical graphics, and data visualization. URL http://www.datavis.ca/milestones (2001)

44. Fruchterman, T.M., Reingold, E.M.: Graph drawing by force-directed placement. Software: Practice and experience **21**(11), 1129–1164 (1991)

45. Gaifman, H.: Dependency systems and phrase-structure systems. Information and control **8**(3), 304–337 (1965)

46. Gretarsson, B., Odonovan, J., Bostandjiev, S., Höllerer, T., Asuncion, A., Newman, D., Smyth, P.: Topicnets: Visual analysis of large text corpora with topic modeling. ACM Transactions on Intelligent Systems and Technology (TIST) **3**(2), 23 (2012)

47. Griffiths, T.L., Steyvers, M.: Finding scientific topics. Proceedings of the National Academy of Sciences **101**(suppl 1), 5228–5235 (2004)

48. HARRIS, J.: Word clouds considered harmful. www.niemanlab.org/2011/10/word-clouds-considered-harmful/ (2011). [Online; accessed Jan. 2016]

49. Harris, Z.S.: Distributional structure. Word **10**(23), 146–162 (1954)

50. Havre, S., Hetzler, E., Whitney, P., Nowell, L.: Themeriver: Visualizing thematic changes in large document collections. Visualization and Computer Graphics, IEEE Transactions on **8**(1), 9–20 (2002)

51. Hays, D.G.: Dependency theory: A formalism and some observations. Language pp. 511–525 (1964)

52. Heintze, N., et al.: Scalable document fingerprinting. In: 1996 USENIX workshop on electronic commerce, vol. 3 (1996)

53. Hilpert, M.: Dynamic visualizations of language change: Motion charts on the basis of bivariate and multivariate data from diachronic corpora. International Journal of Corpus Linguistics **16**(4), 435–461 (2011)

54. Hoad, T.C., Zobel, J.: Methods for Identifying Versioned and Plagiarised Documents. Journal of the ASIS&T **54**, 203–215 (2003). doi:10.1002/asi.10170

55. Holsti, O.R., et al.: Content analysis. The handbook of social psychology **2**, 596–692 (1968)

56. Houvardas, J., Stamatatos, E.: N-gram feature selection for authorship identification. In: Artificial Intelligence: Methodology, Systems, and Applications, pp. 77–86. Springer (2006)
57. Jaffe, A., Naaman, M., Tassa, T., Davis, M.: Generating summaries and visualization for large collections of geo-referenced photographs. In: Proceedings of the 8th ACM international workshop on Multimedia information retrieval, pp. 89–98. ACM (2006)
58. Jankowska, M., Keselj, V., Milios, E.: Relative n-gram signatures: Document visualization at the level of character n-grams. In: Visual Analytics Science and Technology (VAST), 2012 IEEE Conference on, pp. 103–112. IEEE (2012)
59. Keim, D., Oelke, D., et al.: Literature fingerprinting: A new method for visual literary analysis. In: Visual Analytics Science and Technology, 2007. VAST 2007. IEEE Symposium on, pp. 115–122. IEEE (2007)
60. Kim, J.: Causation, nomic subsumption, and the concept of event. The Journal of Philosophy pp. 217–236 (1973)
61. Koh, K., Lee, B., Kim, B., Seo, J.: Maniwordle: Providing flexible control over wordle. Visualization and Computer Graphics, IEEE Transactions on 16(6), 1190–1197 (2010)
62. Krstajić, M., Bertini, E.: Keim, D.a.: Cloudlines: Compact display of event episodes in multiple time-series. IEEE Transactions on Visualization and Computer Graphics 17(12), 2432–2439 (2011). doi:10.1109/TVCG.2011.179
63. Kurby, C.A., Zacks, J.M.: Segmentation in the perception and memory of events. Trends in cognitive sciences 12(2), 72–79 (2008)
64. Landauer, T.K., Foltz, P.W., Laham, D.: An introduction to latent semantic analysis. Discourse processes 25(2–3), 259–284 (1998)
65. Lasswell, H.D.: Describing the contents of communication. Propaganda, communication and public opinion pp. 74–94 (1946)
66. Leavenworth, R.S., Grant, E.L.: Statistical quality control. Tata McGraw-Hill Education (2000)
67. Lee, B., Riche, N.H., Karlson, A.K., Carpendale, S.: Sparkclouds: Visualizing trends in tag clouds. Visualization and Computer Graphics, IEEE Transactions on 16(6), 1182–1189 (2010)
68. Lee, H., Kihm, J., Choo, J., Stasko, J., Park, H.: ivisclustering: An interactive visual document clustering via topic modeling. In: Computer Graphics Forum, vol. 31, pp. 1155–1164. Wiley Online Library (2012)
69. Li, J., Zheng, R., Chen, H.: From fingerprint to writeprint. Communications of the ACM 49(4), 76–82 (2006)
70. Liu, S., Wang, X., Chen, J., Zhu, J., Guo, B.: Topicpanorama: a full picture of relevant topics. In: Visual Analytics Science and Technology (VAST), 2014 IEEE Conference on, pp. 183–192. IEEE (2014)
71. Lotman, I.: The structure of the artistic text
72. Lu, Y., Mei, Q., Zhai, C.: Investigating task performance of probabilistic topic models: an empirical study of plsa and lda. Information Retrieval 14(2), 178–203 (2011)
73. Luo, D., Yang, J., Krstajic, M., Ribarsky, W., Keim, D.: Event river: Visually exploring text collections with temporal references. IEEE Transactions on Visualization and Computer Graphics 18(1), 93–105 (2012). doi:10.1109/TVCG.2010.225. URL http://ieeexplore.ieee.org/xpls/abs_all.jsp?arnumber=5611507
74. Luo, D., Yang, J., Krstajic, M., Ribarsky, W., Keim, D.: Eventriver: Visually exploring text collections with temporal references. Visualization and Computer Graphics, IEEE Transactions on 18(1), 93–105 (2012)
75. Manber, U.: Finding similar files in a large file system. In: 1994 Winter USENIX Technical Conference, vol. 94, pp. 1–10 (1994)
76. Manning, C.D., Raghavan, P., Schütze, H., et al.: Introduction to information retrieval, vol. 1. Cambridge university press Cambridge (2008)
77. Mates, B.: Stoic logic (1953)
78. Mei, Q., Zhai, C.: Discovering evolutionary theme patterns from text: an exploration of temporal text mining. In: Proceedings of the eleventh ACM SIGKDD international conference on Knowledge discovery in data mining, pp. 198–207. ACM (2005)

79. Milgram, S.: Psychological maps of paris, the individual in a social world (1977)
80. Mukhopadhyay, A., Maulik, U., Bandyopadhyay, S.: A survey of multiobjective evolutionary clustering. ACM Computing Surveys (CSUR) **47**(4), 61 (2015)
81. Ng, A.Y., Jordan, M.I., Weiss, Y., et al.: On spectral clustering: Analysis and an algorithm. Advances in neural information processing systems **2**, 849–856 (2002)
82. Oelke, D., Kokkinakis, D., Keim, D.A.: Fingerprint matrices: Uncovering the dynamics of social networks in prose literature **32**(3pt4), 371–380 (2013)
83. Oelke, D., Spretke, D., Stoffel, A., Keim, D.A.: Visual readability analysis: How to make your writings easier to read. Visualization and Computer Graphics, IEEE Transactions on **18**(5), 662–674 (2012)
84. Pirolli, P., Schank, P., Hearst, M., Diehl, C.: Scatter/gather browsing communicates the topic structure of a very large text collection. In: Proceedings of the SIGCHI conference on Human factors in computing systems, pp. 213–220. ACM (1996)
85. Playfair, W.: Commercial and political atlas and statistical breviary (1786)
86. Pylyshyn, Z.W., Storm, R.W.: Tracking multiple independent targets: Evidence for a parallel tracking mechanism*. Spatial vision **3**(3), 179–197 (1988)
87. Ribler, R.L., Abrams, M.: Using visualization to detect plagiarism in computer science classes. In: Proceedings of the IEEE Symposium on Information Vizualization, p. 173. IEEE Computer Society (2000)
88. Riehmann, P., Potthast, M., Stein, B., Froehlich, B.: Visual Assessment of Alleged Plagiarism Cases. Computer Graphics Forum **34**(3), 61–70 (2015). doi:10.1111/cgf.12618. URL http://doi.wiley.com/10.1111/cgf.12618
89. Rivadeneira, a.W., Gruen, D.M., Muller, M.J., Millen, D.R.: Getting Our Head in the Clouds: Toward Evaluation Studies of Tagclouds. 25th SIGCHI Conference on Human Factors in Computing Systems, CHI 2007 pp. 995–998 (2007). doi:10.1145/1240624.1240775
90. Rivadeneira, A.W., Gruen, D.M., Muller, M.J., Millen, D.R.: Getting our head in the clouds: toward evaluation studies of tagclouds. In: Proceedings of the SIGCHI conference on Human factors in computing systems, pp. 995–998. ACM (2007)
91. Robertson, G., Fernandez, R., Fisher, D., Lee, B., Stasko, J.: Effectiveness of animation in trend visualization. Visualization and Computer Graphics, IEEE Transactions on **14**(6), 1325–1332 (2008)
92. Rohrer, R.M., Ebert, D.S., Sibert, J.L.: The shape of shakespeare: visualizing text using implicit surfaces. In: Information Visualization, 1998. Proceedings. IEEE Symposium on, pp. 121–129. IEEE (1998)
93. Seifert, C., Ulbrich, E., Granitzer, M.: Word clouds for efficient document labeling. In: Discovery Science, pp. 292–306. Springer (2011)
94. Sgall, P.: Dependency-based formal description of language. The Encyclopedia of Language and Linguistics **2**, 867–872 (1994)
95. Shivakumar, N., Garcia-Molina, H.: Finding near-replicas of documents on the web. In: The World Wide Web and Databases, pp. 204–212. Springer (1998)
96. Sinclair, J.: Corpus, concordance, collocation. Oxford University Press (1991)
97. Slingsby, A., Dykes, J., Wood, J., Clarke, K.: Interactive tag maps and tag clouds for the multiscale exploration of large spatio-temporal datasets. In: Information Visualization, 2007. IV'07. 11th International Conference, pp. 497–504. IEEE (2007)
98. Smith, A.E., Humphreys, M.S.: Evaluation of unsupervised semantic mapping of natural language with leximancer concept mapping. Behavior Research Methods **38**(2), 262–279 (2006)
99. Steyvers, M., Griffiths, T.: Probabilistic topic models. Handbook of latent semantic analysis **427**(7), 424–440 (2007)
100. Subašić, I., Berendt, B.: Web Mining for Understanding Stories through Graph Visualisation. 2008 Eighth IEEE International Conference on Data Mining pp. 570–579 (2008). doi:10.1109/ICDM.2008.138. URL http://ieeexplore.ieee.org/lpdocs/epic03/wrapper.htm?arnumber=4781152

101. Sun, G., Wu, Y., Liu, S., Peng, T.Q., Zhu, J.J.H., Liang, R.: EvoRiver: Visual Analysis of Topic Coopetition on Social Media. Visualization and Computer Graphics, IEEE Transactions on **PP**(99), 1 (2014). doi:10.1109/TVCG.2014.2346919
102. Svartvik, J.: The Evans statements. University of Goteburg (1968)
103. Teh, Y.W., Jordan, M.I., Beal, M.J., Blei, D.M.: Hierarchical dirichlet processes. Journal of the american statistical association (2012)
104. Tufte, E.R.: Envisioning information. Optometry & Vision Science **68**(4), 322–324 (1991)
105. Tufte, E.R.: Beautiful evidence. New York (2006)
106. Van Ham, F., Wattenberg, M., Viégas, F.B.: Mapping text with phrase nets. IEEE Transactions on Visualization & Computer Graphics **6**, 1169–1176 (2009)
107. Viégas, F.B., Wattenberg, M., Feinberg, J.: Participatory visualization with wordle. IEEE Transactions on Visualization and Computer Graphics **15**(6), 1137–1144 (2009). doi:10.1109/TVCG.2009.171
108. Vuillemot, R., Clement, T., Plaisant, C., Kumar, A.: What's being said near martha? exploring name entities in literary text collections. In: Visual Analytics Science and Technology, 2009. VAST 2009. IEEE Symposium on, pp. 107–114. IEEE (2009)
109. Wang, C., Blei, D., Heckerman, D.: Continuous time dynamic topic models. arXiv preprint arXiv:1206.3298 (2012)
110. Wang, X., McCallum, A.: Topics over time: a non-markov continuous-time model of topical trends. In: Proceedings of the 12th ACM SIGKDD international conference on Knowledge discovery and data mining, pp. 424–433. ACM (2006)
111. Wattenberg, M.: Arc diagrams: visualizing structure in strings. Information Visualization Proceedings **2002**(2002), 110–116 (2002). doi:10.1109/INFVIS.2002.1173155
112. Wattenberg, M., Viégas, F.B.: The word tree, an interactive visual concordance. IEEE Transactions on Visualization and Computer Graphics **14**(6), 1221–1228 (2008). doi:10.1109/TVCG.2008.172
113. Wei, F., Liu, S., Song, Y., Pan, S., Zhou, M.X., Qian, W., Shi, L., Tan, L., Zhang, Q.: Tiara: a visual exploratory text analytic system. In: Proceedings of the 16th ACM SIGKDD international conference on Knowledge discovery and data mining, pp. 153–162. ACM (2010)
114. Werlich, E.: A text grammar of English. Quelle & Meyer (1976)
115. Wu, Y., Provan, T., Wei, F., Liu, S., Ma, K.L.: Semantic-Preserving Word Clouds by Seam Carving. Computer Graphics Forum **30**(3), 741–750 (2011). doi:10.1111/j.1467-8659.2011. 01923.x. URL http://doi.wiley.com/10.1111/j.1467-8659.2011.01923.x
116. Xu, P., Wu, Y., Wei, E., Peng, T.Q., Liu, S., Zhu, J.J.H., Qu, H.: Visual analysis of topic competition on social media. IEEE Transactions on Visualization and Computer Graphics **19**(12), 2012–2021 (2013). doi:10.1109/TVCG.2013.221
117. Xu, T., Zhang, Z., Yu, P.S., Long, B.: Evolutionary clustering by hierarchical dirichlet process with hidden markov state. In: Data Mining, 2008. ICDM'08. Eighth IEEE International Conference on, pp. 658–667. IEEE (2008)
118. Xu, W., Gong, Y.: Document clustering by concept factorization. In: Proceedings of the 27th annual international ACM SIGIR conference on Research and development in information retrieval, pp. 202–209. ACM (2004)
119. Xu, W., Liu, X., Gong, Y.: Document clustering based on non-negative matrix factorization. In: Proceedings of the 26th annual international ACM SIGIR conference on Research and development in informaion retrieval, pp. 267–273. ACM (2003)
120. Zacks, J.M., Tversky, B.: Event structure in perception and conception. Psychological bulletin **127**(1), 3 (2001)
121. Zhang, J., Ghahramani, Z., Yang, Y.: A probabilistic model for online document clustering with application to novelty detection. In: Advances in Neural Information Processing Systems, pp. 1617–1624 (2004)
122. Zhao, Q., Mitra, P.: Event Detection and Visualization for Social Text Streams. Event London pp. 26–28 (2007). URL http://www.icwsm.org/papers/3--Zhao-Mitra.pdf
123. Zheng, R., Li, J., Chen, H., Huang, Z.: A framework for authorship identification of online messages: Writing-style features and classification techniques. Journal of the American Society for Information Science and Technology **57**(3), 378–393 (2006)

Chapter 6
Visualizing Sentiments and Emotions

Abstract Sentiment analysis, also known as opinion mining, is one of the most important text mining tasks and has been widely used for analyzing, for example, reviews or social media data for various of applications, including marketing and customer service. In general, "sentiment analysis aims to determine the attitude of a speaker or a writer with respect to some topic or the overall contextual polarity of a document. The attitude may be his or her judgment or evaluation, affective state (i.e., the emotional state of the author when writing), or the intended emotional communication (i.e., the emotional effect the author wishes to have on the reader)" (Wikipedia, Sentiment analysis—Wikipedia, the free encyclopedia, 2006. https://en. wikipedia.org/wiki/Sentiment_analysis [17]). The sentiment analysis result is usually a score that ranges from -1 to 1 (after normalization) with -1 indicating the most negative, 1 indicating the most positive, and 0 indicating neutral. This sentiment score is usually treated as an attribute of the corresponding text, which can be intuitively differentiated by colors that range from, for example, red (-1) to green (1). In this chapter, we introduce the state-of-the-art sentiments visualization techniques that can be largely classified into two categories including (1) the techniques for summarizing the sentiment dynamics over time, and (2) the techniques for assisting sentiment analysis.

6.1 Introduction

The design goal of a large category of sentiment visualization techniques is to visually summarize the change of sentiments over time regarding to in a given streaming dataset such as news corpus, review comments, and Twitter streams. This goal can be approached, as shown in Fig. 6.1, by showing the sentiment dynamics in a time-series diagram where the time-series curve illustrates the change of sentiment scores computed across the entire dataset at different time points. However this simple visualization is too abstract to display information details such as the causes behind the sentiment shifts. Therefore, many other highly advanced techniques have been introduced to illustrate and interpret the sentiment dynamics from different prospectives.

© Atlantis Press and the author(s) 2016
C. Nan and W. Cui, *Introduction to Text Visualization*, Atlantis Briefs
in Artificial Intelligence 1, DOI 10.2991/978-94-6239-186-4_6

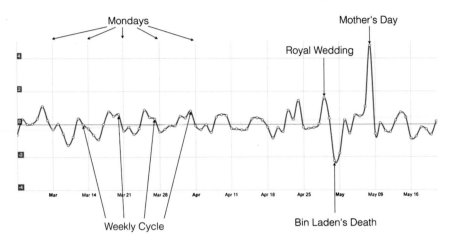

Fig. 6.1 Sentiment indexing of Twitter data in a time-series diagram. This figure shows that the public sentiment regarding different events in real-life may change dramatically

Most of the techniques are developed to compute and visualize the sentiments of a group of focused people based on the text data produced by them. The resulting visualization forms a "happiness indicator" that captures the sentiment change of the focal group over time. For example, Brew et al. [2] introduced SentireCrowds, which represents the sentiment changes of a group of Twitter users from the same city in a timeline view and summarizes the potential underlying event that causes the changes in a multi-level TagCloud designed on the basis of Treemap. Guzman et al. [7] visualizes the change of emotions of groups of different developers in various software development projects. Hao et al. [8] analyzes sentiments through geo-temporal term associations based on a streaming dataset of customers' feedback. Kempter et al. [10] introduced a fine-grained, multi-category emotion model to classify the emotional reactions of users in public events overtime and to visualize the results in a radar diagram, called EmotionWatch, as shown in Fig. 6.3.

Some visual analysis systems have also been developed to support dynamic sentiment analysis. For example, Wanner et al. [15] developed a small multiple visualization view to conduct a semi-automatic sentiment analysis of large news feeds. In this work, a case study on news regarding to the US presidential election in 2008 shows how visualization techniques can help analysts to draw meaningful conclusions without existing effort to read the news content. Brooks et al. [3] introduced Agave, a collaborative visual analysis system for exploring events and sentiment over time in large Twitter datasets. The system employs multiple co-ordinated views in which a streamgraph (Fig. 6.2) is used to summarize the changes of the sentiments of a subset of tweets queried on the basis of user preferences. Zhang et al. [20] introduced a spatial-temporal view for visualizing the sentiment scores of microblog data based on an electron cloud model intruded in physics. The resulting visualization maps a single sentiment score to a position inside a circular visualization display (Fig. 6.3).

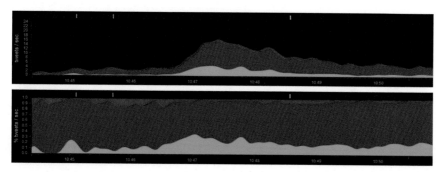

Fig. 6.2 Sentiment streamgraphs for the keyword search Flacco, the Super Bowl MVP in a Twitter dataset using Agave [3]. *Red* indicates negative, *gray* indicates neutral, and *blue* indicates positive. *Top* overall frequency of tweets, divided by sentiment type. *Bottom* sentiment as percentage of overall volume

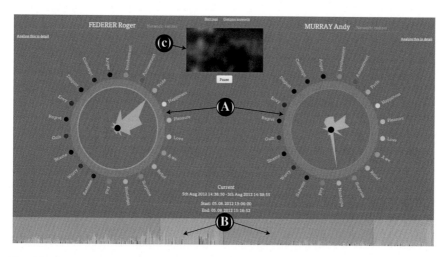

Fig. 6.3 Comparison of two emotion profiles of Roger Federer and Andy Murray (two tennis athletes) after a tennis game in EmotionWatch [10]; (A) the EmotionWatches, (B) timelines showing the two emotion flows, and (C) video

In terms of application, a large set of techniques has been developed to represent the customer's sentiments based on the review data. Alper et al. [1] introduced OpinionBlocks, an interactive visualization tool to enhance understanding of customer reviews. The visualization progressively discloses text information at different granularities from the keywords to the snippets the keywords are used in, and to the reviews containing those snippets. This text information is displayed within two horizontally regions separated by their sentiments. Gamon et al. [6] introduced Pulse for mining topics and sentiment orientation jointly from free text customer feedback. This system enables the exploration of large quantities of customer review data and was used to visually analyze a database of car reviews. Through this sys-

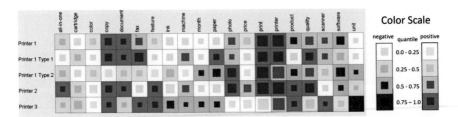

Fig. 6.4 Summary report of printers: each *row* shows the attribute performances of a specific printer. *Blue color* represents comparatively positive user opinions and *red color* comparatively negative ones (see *color scale*). The size of an *inner rectangle* indicates the amount of customers that commented on an attribute. The larger the rectangle the more comments have been provided by the customers

tem, the users can examine customer opinion at a glance or explore the data at a finer level of detail. Oelke et al. [12] analyzed to determine customers' positive and negative opinions via the comments or ratings posted by the customers and visualized the analysis results in a heatmap view showing both volume of comments and the summarized sentiments (Fig. 6.4). More generic systems were also developed. For example, Wensel [16] introduced VIBES, which extracts the important topics from a blog, measures the emotions associated with those topics, and represents topics in the context of emotions based on multiple coordinated views. Makki et al. [11] introduced an interactive visualization to engage the user in the process of polarity assignment to improve the quality of the generated lexicon used for sentiment or emotion analysis via minimal user effort.

Despite the aforementioned techniques in which standard visualizations such as line charts, streaming graphs, and multiple coordinated views are employed. More sophisticated systems with advanced visualization designs were also proposed and developed to analyze the change of sentiments based on the streaming text data. For example, Wang et al. [14] introduced SentiView, which employs advanced sentiment analysis techniques as well as visualization designs to analyze the change of public sentiments regarding popular topics on the Internet. Other systems were designed to analyze sentiment divergences (i.e., conflicting of opinions) that occur between two groups of people. For example, Chen et al. [5] introduced the first work on this topic based on a simple time-series design that summarizes the overall conflicting of opinions based on the Amazon review data. Following this topic, Cao et al. [4] introduced a more advanced technique called SocialHelix, which extracts two groups of people that have the most significant sentiment divergence over time from Twitter data and visualizes their divergence in a Helix visualization as shown in Fig. 6.12, which illustrates how the divergence occurred, evolved, and terminated. Wu et al. [19] introduced OpinionSeer (Fig. 6.5), a visualizaiton design that employs subjective logic [9] to analyze customer opinions about hotels based on their review data inside a simplex space, which is visualized in a triangle surrounded by the context about the customers such as their ages and their countries of origin. Zhao et al. [21] introduced PEARL, which visualizes the change of a person's emotion or mood profiles derived

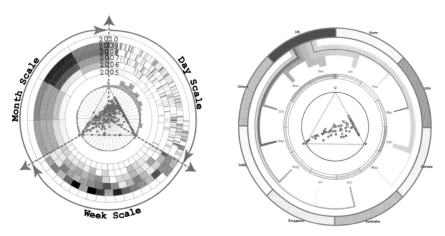

Fig. 6.5 OpinionSeer visualization showing customers' opinions on hotel rooms. The *center trian-gle* is a ternary plot which illustrates the distribution of customers in the three dimensional opinion space computed based on subjective logic. In particular a *dot in the triangle* illustrate a customer, each *triangle vertex* indicates a type of opinion, the distances between the *dot* and *triangle vertex* indicate how strong the customer holds the opinion (i.e., closer is stronger). The surrounding *bar charts* and *arcs* showing context information such as the number of people with similar opinions over time and across different space

from his tweets in a compound belt visualization. The belt groups a set of emotion bands, each one indicating a type of emotion differentiated by colors. The thickness of the band changes over time depending on the portion of the corresponding emotion at different times.

In the rest of this chapter, we describe several well-designed visualizations and the corresponding visual analysis systems that were proposed to reveal and analyze sentiments extracted from various types of text corpus. From these examples, we demonstrate how people's sentiments can be meaningfully visualized and how these sentiment oriented data patterns are revealed.

6.2 Visual Analysis of Customer Comments

In this section, we introduce the visual analysis system developed by Rohrdantz et al. [13] to reveal temporal sentiment patterns in a text stream. The proposed system integrates methods and techniques from text mining, sentiment analysis, and visual analysis together to help analysts to perform a temporal analysis of customers' comments collected from an online web survey. The proposed system helps users to detect interesting portions of an input text streams (e.g., customer reviews/comments), regarding the change of sentiments, data density, and context coherence calculated based on the features (i.e., representative keywords) extracted from the text stream.

In the proposed system, a pixel map is introduced to illustrate the sentiment dynamics of each document (Fig. 6.6) over time. The map provides an overview of the entire corpus by arranging text documents into a pixel based calendar, in which, each pixel indicates a document with the color showing the overall-sentiment of the document. In this case, red indicates negative, green indicates positive, and yellow indicates neutral. The x-axis bins in the background are days and y-axis bins are years with months.

To reveal the change of sentiments, data density, and context coherence, a time density plot is also introduced in the system. As shown in Fig. 6.7, a set of comments that contain the noun "password" (i.e., a keyword feature) extracted from the corpus containing over 50,000 comments, collected over 2 years, are sequentially displayed in a row. In this view, each comment is shown as a vertical bar with color indicating whether the noun password has been mentioned in a positive (green), negative (red) or neutral (gray) context. The height of the bar shows the uncertainty involved in the sentiment analysis following the rule of "the higher the bar, the more certain the result". When hover on a comment bar, the content is displayed in a tooltip. All nouns have background colors to illustrate the sentiments context. The curve plot below these comment bars illustrate the data density over time. An algorithm is also designed in the system to help highlight the data portion that is potentially interesting to the users. With all these visual designs and the pattern detection algorithm, many issues regarding the company have been found from the user comments as shown in Fig. 6.8.

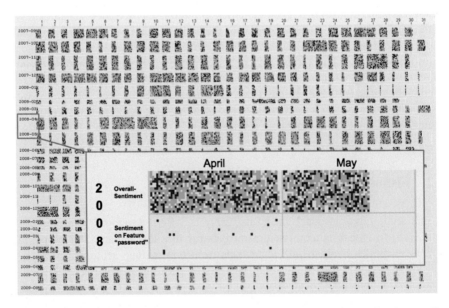

Fig. 6.6 Pixel map calendar: each pixel indicates a document with the color encoding its overall-sentiment, i.e., the average of all contained feature sentiments. Here, *red* indicates negative, *green* indicates positive, and *yellow* indicates neutral. In the background, the x-axis bins are days and y-axis bins are years with months

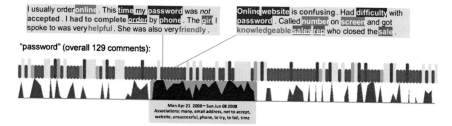

Fig. 6.7 Time density plots visualization that illustrates all the comments related to the feature "password" with associated terms shown at the *bottom* and automatically annotated example comments shown on *top*. Each comment is visualized as a *vertical bar with color* indicates whether the noun password has been mentioned in a positive (*green*), negative (*red*) or neutral (*gray*) context. The height of a bar encodes the uncertainty involved in the sentiment analysis. The *curve* plotted below the comment sequence illustrate the data density over time

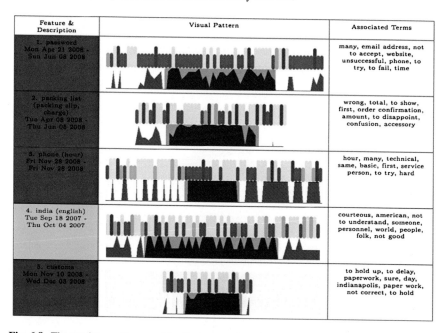

Fig. 6.8 The top issues discovered by the pattern detection algorithm and visualized in the time density plot

6.3 Visualizing Sentiment Diffusion

Understanding and tracing the diffusion process of public opinions has attracted significant attentions in recent years. For example, the government may want to know people's opinion about, for example, a new policy. The sales man in a company may also want to understand how effective a users' opinion will affect others in terms of choosing a product. However, tracing the diffusion of opinion or sentiment is not an easy task because of the rapid spreading and great diversity of public opinions on social media. To address this problem, Wu et al. [18] introduced OpinionFlow, a visual analysis system designed and developed to analyze opinion diffusion on Social Media.

Specifically, the system starts by analyzing a set of tweets posted by users regarding to a predefined topic. For each user u at a given time t, the system computes the sentiment score, denoted as $s(u, t)$, to reveal his/her opinion (either positive or negative) at the moment indicated by t based on the tweets the focal user involved (i.e., post or retweed) in. The system tracks the spreading of opinions by inferring how a single opinion spreads from one user to another by detecting the opinion propagation as a trend formulated by a group of users sharing similar spreading patterns. A statistical analysis model is introduced in the system to estimate the probability that an opinion initiated by user u would propagate to another user v given a predefined topic. The diffusion under a group of multiple topics is calculated and illustrated in a flow-based visualization simultaneously.

In particular, OpinionFlow employs a Sankey diagram based design to represent user flows across multiple topics because of its simplicity and the intuitiveness of

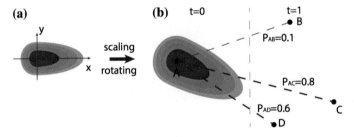

Fig. 6.9 The **a** adapted and **b** rotated Gaussian kernel function used for representing the direction and density for representing the opinion diffusion process

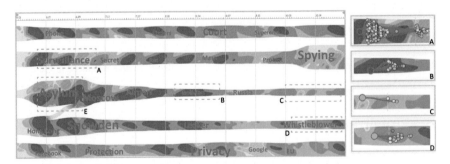

Fig. 6.10 The visualization of PRISM data based on Opinion Flow which indicates five major topics. *A* and *D* illustrate the spreading paths of the media user "FoxNews" with strong and weak opinions. *B* and *C* illustrate the diffusion paths of a common user with strong and weak opinions, respectively

representing information flows. On top of this view, a density map is employed to help illustrate the diffusion of opinions among users over time. Specifically in order to reveal the diffusion trend, an altered Gaussian kernel is used as shown in Fig. 6.9. The resulting visualization reveals many interesting diffusion pattern as shown in Fig. 6.10. In this view, a node-link diagram is also used to specifically highlight the individual diffusion path on top of the density map, showing more details regarding the diffusion.

6.4 Visualizing Sentiment Divergence in Social Media

SocialHelix [4] is the third example introduced in this chapter to demonstrate the usefulness of visualization techniques in terms of revealing sentiment oriented data patterns. SocialHelix was designed to help uncover the divergence of users' opinions (i.e., sentiment divergence) inspired by an event in social media platforms such as Twitter. For example, in political campaigns, people who support different parties and candidates may engage in debates on social media based on their own political perspectives. Makers of competing products may also launch persuasion campaigns on social media to attract attention. From these examples, we can see that sentiment divergence is a complex social phenomenon that involves three key components, i.e., the communities of users who are involved in the divergence by holding different opinions, the topic context in which the divergence is occurred and developed, as well as various events or episodes that provoke the divergence. In this case, several primary characteristics of the components can be observed. For example, a divergence should at least involve two groups of people (i.e., communities) and these groups must hold different opinions regarding the same series of focal events or episodes over time. These components as well as the corresponding characteristics are used to detect divergence from the raw social media data and also to design visualizations to represent the divergence.

Specifically, the SocialHelix system detects and represents the temporal process of social divergence through a series of data processing steps. With a given set of streaming Twitter data, the SocialHelix system first extracts a group of active users who continuously post or retweeted a large number of tweets over time. In the following, the system computes their sentiments at different time points based on their posts. The results form a vector $S = [s_1, s_2, \ldots, s_n]$ where s_i is a numerical score that indicates the users' sentiment at the i-th time point. This vector is then used for cluster analysis to group users with similar sentiment changing trend together. Among all the resulting clusters, the two with the most significant differences (i.e., with the largest sentiment divergence) are selected to be visualized in a novel helix based visualization design as shown in Fig. 6.11. This design is inspired by the structure of DNA helix given its perfect match with the aforementioned key components in a typical sentiment divergence procedure. In particular, in this design, the backbones of the DNA helix are used to illustrate two communities with the largest sentiment

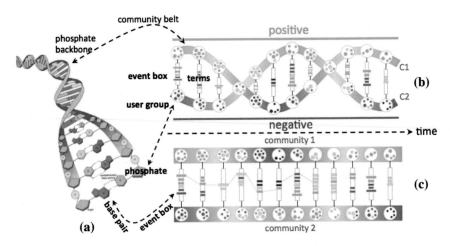

Fig. 6.11 The design of SocialHelix visualization employs a DNA helix-metaphor in which the backbones of the DNA helix is used to illiterate two communities with the largest sentiment divergence. The phosphates represent the groups of representative users in these communities at different time and the base pairs are used to illustrate the episode or events that inspired the divergence

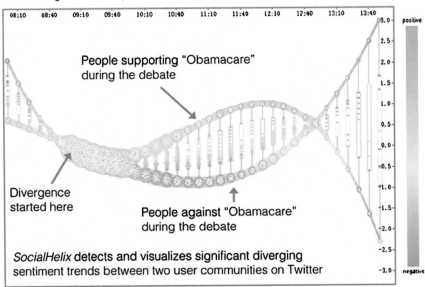

Fig. 6.12 SocialHelix visualization summarizes the sentiment divergence of two group of Twitter users based on the sentiment analysis of their posts. This figure illustrates the divergence about "Obamacare" between two groups of people during the 2012 US presidential debate on October 3rd, 2012

divergence. The phosphates represent the groups of representative users in these communities at different times. The base pairs are used to illustrate the episodes or events that inspired the divergence.

Based on a Twitter dataset collected during four political debates during the 2012 presidential election in the US, the SocialHelix system revealed many interesting findings. One example is shown in Fig. 6.12. This figure illustrates a divergence between two groups of people centered around the topic of "ObamaCare", a new healthcare policy in USA proposed by President Obama. One group of users shows support but another is against this policy. The data were collected during the live debate show on TV occurred in the evening of Oct 3rd, 2012. This figure clearly illustrates the divergence started after the political debate (i.e. after 10:00 p.m.), when the viewers of the debate started to post their opinions on Twitter. An interesting finding is the emergence of turning point in which the group of users holding negative opinions shifted to positive while the group with positive opinions shifted to negative, which was suspicious point and worth a close investigation of the detailed tweets.

6.5 Conclusion

This chapter has introduced sentiment visualization techniques. After briefly reviewing the existing visualization and visual analysis techniques, we introduced three example systems and the corresponding visualization designs to help readers understand how visualization can help with the representation and analysis of sentiment-oriented information. Sentiment visualization is a highly interesting research topic that is under development. Although we have introduced many techniques in this chapter, challenges such as accuracy and scalability issues have not been fully addressed. In terms of visualization, a systematic approach or a well-established standard for representing sentiment-oriented information is lacking. The current visualization techniques are more application-driven and cannot be clearly categorized. Therefore, we believe that further study in this direction is necessary and promising.

References

1. Alper, B., Yang, H., Haber, E., Kandogan, E.: Opinionblocks: visualizing consumer reviews. In: IEEE VisWeek 2011 Workshop on Interactive Visual Text Analytics for Decision Making (2011)
2. Brew, A., Greene, D., Archambault, D., Cunningham, P.: Deriving insights from national happiness indices. In: 2011 IEEE 11th International Conference on Data Mining Workshops (ICDMW), pp. 53–60. IEEE (2011)
3. Brooks, M., Robinson, J.J., Torkildson, M.K., Aragon, C.R., et al.: Collaborative visual analysis of sentiment in twitter events. In: Cooperative Design, Visualization, and Engineering, pp. 1–8. Springer, Berlin (2014)
4. Cao, N., Lu, L., Lin, Y.R., Wang, F., Wen, Z.: Socialhelix: visual analysis of sentiment divergence in social media. J. Vis. **18**(2), 221–235 (2015)

5. Chen, C., Ibekwe-SanJuan, F., SanJuan, E., Weaver, C.: Visual analysis of conflicting opinions. In: 2006 IEEE Symposium on Visual Analytics Science and Technology, pp. 59–66. IEEE (2006)
6. Gamon, M., Aue, A., Corston-Oliver, S., Ringger, E.: Pulse: Mining customer opinions from free text. In: Advances in Intelligent Data Analysis VI, pp. 121–132. Springer, Berlin (2005)
7. Guzman, E.: Visualizing emotions in software development projects. In: IEEE Working Conference on Software Visualization, pp. 1–4. IEEE (2013)
8. Hao, M.C., Rohrdantz, C., Janetzko, H., Keim, D.A., et al.: Visual sentiment analysis of customer feedback streams using geo-temporal term associations. Inf. Vis. **12**(3–4), 273 (2013)
9. Jøsang, A.: The consensus operator for combining beliefs. Artif. Intell. **141**(1), 157–170 (2002)
10. Kempter, R., Sintsova, V., Musat, C., Pu, P.: Emotionwatch: visualizing fine-grained emotions in event-related tweets. In: International AAAI Conference on Weblogs and Social Media (2014)
11. Makki, R., Brooks, S., Milios, E.E.: Context-specific sentiment lexicon expansion via minimal user interaction. In: Proceedings of the International Conference on Information Visualization Theory and Applications (IVAPP), pp. 178–186 (2014)
12. Oelke, D., Hao, M., Rohrdantz, C., Keim, D., Dayal, U., Haug, L.E., Janetzko, H., et al.: Visual opinion analysis of customer feedback data. In: IEEE Symposium on Visual Analytics Science and Technology, 2009. VAST 2009, pp. 187–194. IEEE (2009)
13. Rohrdantz, C., Hao, M.C., Dayal, U., Haug, L.E., Keim, D.A.: Feature-based visual sentiment analysis of text document streams. ACM Trans. Intell. Syst. Technol. (TIST) **3**(2), 26 (2012)
14. Wang, C., Xiao, Z., Liu, Y., Xu, Y., Zhou, A., Zhang, K.: Sentiview: sentiment analysis and visualization for internet popular topics. IEEE Trans. Hum. Mach. Syst. **43**(6), 620–630 (2013)
15. Wanner, F., Rohrdantz, C., Mansmann, F., Oelke, D., Keim, D.A.: Visual sentiment analysis of rss news feeds featuring the us presidential election in 2008. In: Workshop on Visual Interfaces to the Social and the Semantic Web (VISSW) (2009)
16. Wensel, A.M., Sood, S.O.: Vibes: visualizing changing emotional states in personal stories. In: Proceedings of the 2nd ACM International Workshop on Story Representation, Mechanism and Context, pp. 49–56. ACM (2008)
17. Wikipedia: Sentiment analysis—Wikipedia, the free encyclopedia (2006). https://en.wikipedia.org/wiki/Sentiment_analysis. Accessed 10 Nov 2015
18. Wu, Y., Liu, S., Yan, K., Liu, M., Wu, F.: Opinionflow: visual analysis of opinion diffusion on social media. IEEE Trans. Vis. Comput. Graph. **20**(12), 1763–1772 (2014)
19. Wu, Y., Wei, F., Liu, S., Au, N., Cui, W., Zhou, H., Qu, H.: Opinionseer: interactive visualization of hotel customer feedback. IEEE Trans. Vis. Comput. Graph. **16**(6), 1109–1118 (2010)
20. Zhang, C., Liu, Y., Wang, C.: Time-space varying visual analysis of micro-blog sentiment. In: Proceedings of the 6th International Symposium on Visual Information Communication and Interaction, pp. 64–71. ACM (2013)
21. Zhao, J., Gou, L., Wang, F., Zhou, M.: Pearl: an interactive visual analytic tool for understanding personal emotion style derived from social media. In: 2014 IEEE Conference on Visual Analytics Science and Technology (VAST), pp. 203–212. IEEE (2014)